ALSO BY DAN RHODES

Anthropology

'A book you'll want to hurl from rooftops at passers-by to spread the word.'
UNCUT

'A gleaming box of jazzy miniatures. Exquisitely funny.'
GUARDIAN

'Touching and insightful . . . you'll want to devour every one.'
HEAT

'Effortless to read, amusing and yet coloured by a deep sadness about the passing of things, you will want to hold on to the truths it so skilfully offers for as long as you would to love.'
INDEPENDENT

'Very funny and very sharp.'
THE TIMES

'I cannot express to you how much this book delighted me. Go and read it.'
BIG ISSUE

'Anybody who liked *There's Something About Mary* will love Rhodes's book, and that's a strange thing to be able to say about a piece of conceptual art.'
LA WEEKLY

'A great collection, by turns funny, dark and touching.'
Dave Gorman

Don't Tell Me The Truth About Love

'*Anthropology* was both funny and unsettling. *Don't Tell Me The Truth About Love* carries on from there. These are fairy tales written in the detached, ethereal tones of the Brothers Grimm. Needless to say, none of these cautionary tales ends happily ever after, but Rhodes's narrative skill, his clear diagrams and his wrong-footing humour sweeten the pill immeasurably.'

DAILY TELEGRAPH

'A stunning collection . . . Buy it for anyone you care about, including yourself.'

HEAT

'You won't find a finer collection of short love stories anywhere in the land. One of those rare collections that make the heart sing.'

***** JOCKEY SLUT

'Is there a more innovative, but also intensely readable, writer than Dan Rhodes working in Britain today?'

BIG ISSUE

'It is brave of Rhodes to buck the trend for realism . . . the beauty of his writing is persuasive and his themes are universal.'

THE TIMES

Timoleon Vieta Come Home

Winner of the Author's Club First Novel Award
Winner of the QPB New Voices Award
Shortlisted for Le Prince Maurice Prize
Shortlisted for the John Llewellyn Rhys Prize

'Surely the true best of Granta's new Best Of list. Everybody should go out and buy *Timoleon Vieta Come Home*, a tender but unsentimental novel about a failed composer, his sadistic lover and his mongrel dog. A story worthy of W.G. Sebald, universal in its scope and ambition.'
Rose Tremain, DAILY TELEGRAPH

'A dog, a beautiful mongrel, is the hero of Dan Rhodes's first novel, *Timoleon Vieta Come Home*, which is by turns hilarious and heartrending. Rhodes is that real, rare thing – a natural storyteller.'
Paul Bailey, SUNDAY TIMES

'A delight, a masterpiece of beautifully unforced comedy.'
OBSERVER

'I loved it; it is quirky and original, and the storytelling is truly virtuoso. A literary treasure.'
Louis De Bernières

'A tragicomedy heavy on the comedy, *Timoleon Vieta* is an extremely fresh and sensitive meditation on love lost and unresolved anger. A beautiful and often touching book.'
INDEPENDENT ON SUNDAY

'Extraordinary. I haven't read anything like it before. It's a seemingly unemotive but beautifully crafted novel with a big emotional hook at the end. It really smacks you in the face.'
DBC Pierre, GUARDIAN

'Beguiling and affecting . . . an amusing and exhilarating ragbag. I have to say that I rather loved it.'
INDEPENDENT

'A heartbreaking tale of loneliness, longing, betrayal and dogged devotion.'
HERALD

'A novel that's as unusual as it is unforgettable.'
ARENA

'Savagely funny, startlingly original.'
THE TIMES

'Charming, original, funny, biting and wise.'
GUARDIAN

'Part shaggy-dog tale, part fairy-tale, part Lassie take-off, and a quite thoroughly original debut . . . A beguiling and resonant little novel.'
Michiko Kakutani, NEW YORK TIMES

The Little White Car

(writing as Danuta de Rhodes)

'Excellent . . . Beautifully written and very funny.'
INDEPENDENT ON SUNDAY

'Observant, absurd, sentimental and very, very funny.'
DAILY TELEGRAPH

'A cracking good read . . . enjoy yourself.'
DAILY EXPRESS

'Fabulous . . . A very funny, very charming novel that we quite simply insist you read. Danuta is a very talented storyteller indeed.'
***** HEAT

'Wonderfully fresh and witty.'
OBSERVER

'Easy to read, neat and without a hair out of place, *The Little White Car* is an absolute pleasure.'
LIST

'Breezy, funny and charming.'
THE TIMES

'A brilliant (supposed) debut.'
i-D

'A lean, mean hilarity machine.'
ENTERTAINMENT WEEKLY

DAN RHODES

CANONGATE
Edinburgh · London · New York · Melbourne

Acknowledgements

Thanks to Stan, Pru, Fran and Emmily for wading through tatty incarnations of the book. Also to Jamie, Francis and everybody else at Canongate, foreign publishers past and present who have been reckless enough to take a punt on my stuff, the girls in the skyscraper, all the family (from East Leake to Alabang), and to my vicious thug of a literary agent (me).

Special thanks to Judge Ginge, who waded above and beyond the call of duty, and to the pubs and paths of Pembrokeshire.

First published in Great Britain in 2007 by
Canongate Books Ltd, 14 High Street,
Edinburgh EH1 1TE

This paperback edition first published in 2008 by Canongate Books

British Library Cataloguing-in-Publication Data
A catalogue record for this book is available on
request from the British Library

ISBN 978 1 84767 048 9

Typeset in Plantin Light
by Palimpsest Book Production Ltd, Grangemouth, Stirlingshire
Printed and bound in Great Britain by
Clays Ltd, St Ives plc
www.canongate.net

To Wife-features

MONDAY

Tall Mr Hughes, short Mr Hughes and Mr Puw were standing at the bar of The Anchor. 'You know what we would be doing right now if we were alligators?' asked tall Mr Hughes, who had hardly spoken about anything but alligators for three consecutive evenings. It had been *alligators* this, and *alligators* that.

'No,' mumbled Mr Puw, looking, but at the same time not looking, at a row of horse brasses on the far side of the room.

Short Mr Hughes was staring in the direction of the pewter mugs that were hanging from one of the beams behind the bar. He didn't say anything, but he turned down the corners of his mouth and shook his head.

Tall Mr Hughes drew himself up to his full height, even allowing his heels to rise a little off the ground. Despite his name he wasn't particularly tall, just a bit taller than short Mr Hughes, who wasn't particularly short. They were each just an inch or two either side of average, and it was Mr Puw, with his pipe and his big black beard, who was the shortest of the three by some distance.

Deciding he had kept them in suspense for long enough, tall Mr Hughes finally revealed his latest alligator fact. 'We would be . . .' he said, '. . . hibernating.'

'Oh,' said Mr Puw, his eyes at last focusing on the horse brasses. They seemed particularly shiny, and he half-wondered whether they had been polished since the last time he had paid attention to them. He had no idea when that would have been.

Short Mr Hughes carried on staring in the direction of the pewter mugs. He turned down the corners of his mouth and nodded, running the fingers of his left hand over his bristly grey moustache.

'It's winter, you see,' clarified tall Mr Hughes, his rich baritone filling the room, 'and they hibernate in the winter . . .' He stared at his drink for a while, and when he spoke again his voice was quieter. '. . . do alligators.' He lowered his heels, and reached for his drink.

The silence that followed was broken when the bottle refrigerator's thermostat clicked, and sent it rattling and rumbling into life. It was a lot louder than it ought to have been.

'That fridge needs looking at,' said short Mr Hughes.

Tall Mr Hughes and Mr Puw nodded, but neither made

any steps to move the conversation away from alligators and towards refrigerator maintenance.

These had been long evenings. Sometimes tall Mr Hughes would state the obvious, telling them that alligators are, at least for all intents and purposes, carnivores, or that even though they were crocodilian they weren't *actual* crocodiles. Other times he cleared up grey areas, like whether or not they laid eggs. Short Mr Hughes and Mr Puw were fairly sure that alligators *did* lay eggs, but neither of them would have put money on it, mainly because they weren't gambling men but also because they weren't particularly bothered either way. They had had more than their fill of the subject. Their glazed eyes betrayed this, but tall Mr Hughes didn't seem to notice. Drifting back into consciousness, he carried on where he had left off.

'And come the spring,' he said, over the erratic thrum of the refrigerator, 'we would start to become . . .' He paused before the grand revelation, and drew himself up to his full height before looking around to make sure there weren't any women within earshot. The three of them were still the only customers and the barmaid had the night off, leaving the landlord to serve the drinks alone, but even so, tall Mr Hughes leaned in close to the others and continued confidentially, '. . . we would start to become frisky. We would pay special attention to the lady alligators, if you catch my drift. That's what happens in the springtime to our plantigrade pals.'

Mr Puw and short Mr Hughes said nothing, but they both nodded very slightly.

'I suppose you're wondering what *plantigrade* means, aren't you?'

Neither of them was wondering any such thing. They were just trying, without a great deal of success, not to imagine alligators making passionate love. They were both content for him to carry on though, because it was Monday and a quarter past eight, and the quiz was due to start at half past so he would be obliged to stop talking about alligators before too long. They went back to their pints of bitter as the uninvited definition of *plantigrade* washed over them.

'They're just like us, really,' concluded tall Mr Hughes.

'Oh yes,' mumbled Mr Puw. 'The similarities are endless.' His gaze at last shifted away from the horse brasses, and he raised his pipe and nodded a greeting across the bar to Septic Barry and the Children from Previous Relationships as they walked in through the far door, bringing with them a cold blast of the outside air.

'Ready to be humiliated?' called Septic Barry, returning the nod.

'No, but I hope *you* are,' said Mr Puw. 'We'll be wiping the floor with you lot tonight. We have a secret weapon.'

'Oh yes? What's that then – your beard?'

'No, it's not my beard.' Mr Puw was very proud of his big black beard. He was also proud of his big round belly, pure Brains bitter he claimed, and of the way he smoked nothing but pipe tobacco in the age of cigarettes. 'It's something else entirely.'

'Like what?'

'That would be telling, wouldn't it?'

Septic Barry knew, as did everybody else, that they didn't have any kind of secret weapon.

The Children from Previous Relationships hung up their coats, and sat at their usual table on the public bar side of the pub as Septic Barry got the first round in from Mr Edwards, the landlord, who had emerged from his sanctuary behind the door marked PRIVATE and begun pulling four pints of Brains without having to be asked.

The Anchor had two glass-panelled front doors, one engraved with the words LOUNGE BAR and the other with the words PUBLIC BAR, but the days of different prices and different service had gone a long time ago, and the dividing wall had been taken out in the Seventies. Now it was just the same pub all the way through, but each side was still referred to by its old name, and short Mr Hughes, tall Mr Hughes and Mr Puw always came in through the LOUNGE BAR door and stayed on that side of the pub, and Septic Barry and the Children from Previous Relationships always came in through the PUBLIC BAR door and stayed on that side of the pub.

Quite often in the winter it would just be the seven of them, plus whoever was behind the bar, and on the very rare nights when Septic Barry and the Children from Previous Relationships were to spend a whole evening playing pool in the pub on the other side of the harbour, it would just be tall Mr Hughes, short Mr Hughes and Mr Puw. In the summer though, the village was crowded with holidaymakers, and The Anchor would be packed. A lot of the time it was hard to find a place to sit or stand, and although Septic Barry and the Children from Previous Relationships would never miss a night's drinking, the older men found themselves inclined to spend their free time looking after their vegetable patches, or attending to outstanding household projects. Being married, short Mr Hughes and Mr Puw had even been known to spend evenings with their wives. On Mondays they would turn up for quiz night no matter what, getting there early to be sure of a table, but most nights they bided their time, quietly longing for the end of the season when The Anchor would at last become theirs again. Once they had reclaimed it, it was a rare shift that didn't see the three of them standing together at the bar for the best part of the evening.

Their glasses were almost empty.

'Your round, Mr Puw,' said short Mr Hughes.

Mr Puw patted his belly, and said, 'I might have filled out a bit over the years, but there's no need to be personal.'

They had been having this exchange, word for word, for years, but they smiled as Mr Puw drained his final mouthful and put his empty glass on the bar. They all watched Mr Edwards as he handed Septic Barry his change and moved over to their side of the pub, where he pulled three pints of Brains without having to be asked.

Mr Edwards put the change from Mr Puw's ten-pound note on the bar as the three men drank the tops of their new pints, and wiped the froth from their lips with their sleeves. The refrigerator had stopped its rumbling, and there was a silence between them. Tall Mr Hughes decided to fill it. 'As a matter of fact,' he said, 'if you look *really* closely, something you'll always notice about alligators is . . .'

He was cut short when the lounge bar's door opened. It didn't open very far, but far enough for Miyuki Woodward to get through. Everybody turned to look at her, and the place went quiet as they all realised it was that time of year again.

The quiet didn't last long. 'Back from the dead?' called Septic Barry, from his seat on the other side of the pub.

'Back to haunt you,' she said.

'It's the proverbial bad penny,' said tall Mr Hughes, who didn't seem too upset at having been interrupted.

'That's right,' she said. 'You can't get rid of me.'

'Very honourable to see you again,' said short Mr Hughes, as she had known he would.

'Very honourable to see you too,' she sighed.

'Welcome back, Thunderthighs,' said Mr Puw, smiling warmly through his big black beard as he realised that maybe they would have a secret weapon after all. 'You've been sorely missed, you know.'

'That's good to hear.'

Mr Edwards smiled in recognition of her, shook his head, and said, 'Holy mackerel.'

She smiled back. 'The usual, please.' This was a test, because it had been eleven and a half months.

'Holy mackerel,' he said again, chuckling as he reached for a pint glass and filled it with Brains.

He put the glass in front of her, and she gave him a five-pound note. She took her change, thanked him, looked at the coins and calculated that a pint had gone up by five pence, which was no more or less than she would have expected. She picked up the glass with both hands, and drank the top half-inch of the beer.

'See you later then,' she said to tall Mr Hughes, short Mr Hughes and Mr Puw.

'Oh, you'll be seeing us later, Thunderthighs,' said Mr Puw. 'Don't you worry about that.'

She took off her coat, draped it over a stool and sat at the small round table in the corner of the room, underneath the stuffed pike in its glass case. She drank almost half her pint in one go, and looked around. It was as if the last fifty weeks had never happened. The people were the same as when she was last there, and the pub hadn't changed at all. Septic Barry's hair was still short on top and long at the back, coal and wood burned in the lounge bar's fireplace, and even though the dates fell on different days of the week, the January page of the brewery calendar was decorated with the same watercolour dray horse as the year before.

As she felt herself blend into the scene it struck her that she hadn't changed a great deal either. She was drinking what she always drank, and sitting where she always sat, at the end of the bench that ran along the wall from the back corner to the fireplace. She was wearing the same brown jumper and walking boots she had habitually worn to the pub the year before, and although her jeans were technically different from last year's they were from the same shop, they were exactly the same shade of blue, and they were just as spattered with multi-coloured spots of paint. Her hair was more or less the same shortish style as it had been for years, and it was pulled back from her face with a black band that she couldn't remember having bought in the intervening months, and even though she was always resolving to mend her ways, her nails were still bitten down to the quick.

Precisely on cue, a horrible screech of feedback made everybody wince. The equipment had been plugged in, and tall Mr Hughes, short Mr Hughes and Mr Puw took their usual quiz night seats at the other end of the long bench from Miyuki, around the table by the fire.

She left her book unopened in front of her. She knew from experience that she wouldn't be able to concentrate, but it didn't bother her because she had read most of it on the train and bus, and only had a few pages left to go. And besides, she was sure she knew who the murderer was.

* * *

The quizmaster, a retired telephone engineer who was paid in pints of bitter and packets of pork scratchings, had appeared with moments to spare. After turning on his amplifier and hurriedly handing out paper and pens to the contestants, he started talking into his distorted microphone. He gave a brief summary of the rules and the structure of the competition, and began the first round of questions.

With his coarse russet sweep, tatty beige cardigan and relentlessly monotonous voice, there was nothing of the game show host about him, but even so he felt his role as Master of Ceremonies demanded that he inject a little pizzazz into the proceedings, which he chose to do by ending every question with the sound, *Ah*.

How many chains are there in a furlong? became, *How many chains are there in a furlong-ah?* Likewise, *For what do the construction-related initials J.C.B. stand?* was posed as, *For what do the construction-related initials J.C.B. stand-ah?* With his questions held so close to his face that they almost touched his eyeballs, and his delivery suggesting he still had difficulty reading his own handwriting, contestants were asked, *What was the name of the first dog to be fired into orbit-ah?* or *What was the final tally of unfortunate young men to be callously slaughtered by the London-based serial killer Dennis Nilsen-ah?*

Tall Mr Hughes, short Mr Hughes and Mr Puw huddled around their table on the lounge bar side of the pub, and Septic Barry and the Children from Previous

Relationships huddled in a haze of smoke around their table on the public bar side of the pub. To the dismay of both regular teams, four walkers from Usk who were staying in the rooms on the top floor had decided to join in at the last minute. They called themselves The Four Walkers from Usk, and they seemed unnervingly keen as they huddled around the table in between the dartboard and the cigarette machine.

Miyuki played along in her head, knowing some of the answers but not others, and it wasn't until the second question in the fourth round that tall Mr Hughes, short Mr Hughes and Mr Puw remembered she was there. As one, they turned and looked at her. The question was: *What is the new name of the country that used to be called Burma-ah?* Miyuki pretended they weren't all looking at her, but she knew they were. This time it was short Mr Hughes who approached her. Still sitting, he inched along the bench with his pint glass in his hand, smiling in a way that didn't suit him, and before he could say a word she stage-whispered, 'Myanmar.' She quietly spelled it for him.

To make sure he had got it right he repeated the spelling of *Myanmar* back to her, and when she nodded her approval he winked and whispered, 'Very honourable of you,' and slid back along the bench. He passed on the information to his team mates, and tall Mr Hughes wrote the answer on their sheet.

* * *

Something very similar to this had happened on her first evening in the village, eight years before. She had been sitting quietly on her own, underneath the stuffed pike, and trying to read her book as quiz night went on around her, when she felt a presence by her side. That time it had been Mr Puw, with his big round belly and his big black beard, who had slid along the bench.

'We were wondering . . .' he said, slowly and deliberately, nodding his head sideways and pointing his pipe in the direction of his team mates, '. . . us lot . . . whether or not you happened to know the answer to that last question.'

Miyuki said, 'Two hundred and seventy-seven miles per hour.'

'Are you sure about that?'

'I'm positive.'

'That's really fast, though.'

'You don't have to believe me if you don't want to,' she smiled. 'It's up to you.'

'But are you *really* sure?'

She gave him a look.

'Of course you're sure,' he said. 'Silly question.'

Mr Puw's wife's friend handled the changeovers at the cottage Miyuki was renting, and she had told Mr Puw's wife that they were expecting the arrival of somebody with a Japanese-sounding name. It was rare enough for there to be a tenant at this time of year, and rarer still for a tenant to have a Japanese-sounding name, so the women deemed this information to be worthy of mention. The news had been relayed to a half-listening Mr Puw over that morning's breakfast table, and he had correctly deduced that the person sitting at the other end of the long bench was the one his wife had been talking about.

'It is fast though, isn't it?' he said. 'Thanks for that, Thunderthighs.' He slid back to their table, shaking his head at the thought of something so long and so heavy travelling at such a speed, and leaving Miyuki feeling self-conscious about the size of her legs. She had never thought of them as being particularly big before. She had thought they were just normal.

There was a time when she would have been exasperated at being asked whether or not she knew the highest recorded speed of the bullet train to within ten miles per hour either way. She had found it grating when people assumed she knew everything about Japan, and that she would be glad to talk about the country, to discuss its customs and its culture at great length. It had happened a lot when she was at college, and she often ended up telling people that although her father was Japanese she had never met him, and that her mother was Welsh and had moved back home when she was five months pregnant. She hadn't spent any time in Japan since she had been in the womb. 'I was an Osaka foetus,' she would say, as her inquisitor shrank with embarrassment, 'and that's all. I'm about as Japanese as laverbread.'

Gradually, though, she found herself starting to acknowledge that she felt a small connection to the place. Even though her parents had parted company before she was born, and she knew almost nothing about her father, her mother had given her a Japanese name and it was her father's genes that were in the ascendant, at least when it came to her appearance. She had Japanese eyes, a Japanese build and Japanese colouring. She didn't look Korean or Filipino or Chinese or Vietnamese, and she didn't look indeterminately east Asian. There was no escaping it: she looked Japanese.

At college Japanese students had been known to approach her and speak to her in their own language, and she would say, 'I'm very sorry, but I don't know what you're talking about.' After a short and confused conversation she would watch them walk away, and wonder, just for a moment, whether they had been a cousin, or even a half-brother or -sister.

Over time, she began to sympathise with her interrogators. She came to the conclusion that if people wanted to talk to her about Japan then there was no reason why they shouldn't. She had grown to realise that everybody is saddled with the curse of small talk in one way or another. Veterinary assistants trying to relax in general company are tormented with interminable true stories of decrepit parrots, crippled badgers, and poodles with weeping sores; off-duty plumbers trying to wind down in pubs are pestered by fellow drinkers with extensive inquiries about float valves and stopcocks; and undertakers, on revealing their profession to new acquaintances, are obliged to endure bowed heads and reverent hushes followed by horribly sincere mutterings about how it's all such a terrible business when somebody passes away. She was by no means alone, and sometimes she shrivelled with shame as she found herself asking people very obvious questions that they must have been asked all the time. Her insides shrank and her toes wriggled with embarrassment whenever she realised that in clutching at conversational straws she had crossed over to the other side and the stultifying conversation was her fault.

When she read, to her chagrin, of the existence of a Japanese dish that was more or less identical to laverbread, she knew that the time had come at last for her to accept her millstone, and to find out at least some of the answers to the questions she was so often asked. She started watching Japanese arthouse films, horror movies and anime, and television documentaries about life in Tokyo, and geishas, and the aftermath of Hiroshima. She borrowed travel guides and coffee-table books from the library, and asked her mother for manga comics for Christmas, and read books by Haruki Murakami, Yukio Mishima and Banana Yoshimoto. She ended up amassing quite an extensive range of knowledge, and even found that her long-standing and carefully cultivated indifference towards the country had been replaced by a genuine moderate interest, so when people asked her questions about things to do with Japan she no longer expended energy being affronted, but instead battered them with basic historical, cultural and geographical information. If she didn't know the answer she would make a point of finding it out and getting back to them. She learned the height of Mount Fuji, the names of the big cities and the main islands, the primary components of a sumo wrestler's diet, the ideal temperature of sake and the numbers up to a hundred, as well as a raft of examples of Japanese etiquette and details about the preparation and consumption of puffer fish. Until then the only trains that had ever really captured her attention had been the handsome, snub-nosed Class 37s as they rattled up her Valley on their way to the colliery and, laden with coal, defiantly lumbered their way back down, but she memorised every fact she could find about the bullet train.

Because people's curiosity didn't stop at the edge of Japanese waters, she stocked up on all kinds of east Asian trivia, which she dispensed with casual generosity. She knew every capital city in the region, the length of the Great Wall of China and the dates of its construction, the name of the official language of the Philippines, the capital of Taiwan, the key differences between Malaysia and Indonesia, and, helpfully for the two Mr Hugheses and Mr Puw, the new name for the country that used to be called Burma. When people asked her, as they often did, whether it was true that you could buy schoolgirls' used knickers from vending machines at Tokyo subway stations, they would be told to go away and not come back until they had a sensible question, but by and large she dealt with all inquiries as thoroughly, and as politely, as she could.

On her first night in the pub her answer about the bullet train, although absolutely correct, had been disallowed. From the far side of the room Septic Barry had spotted Mr Puw's manoeuvre, and complained to the quizmaster about their team receiving illegal help from a third party. It was a valid grievance and the point was deducted, but tall Mr Hughes, short Mr Hughes and Mr Puw had since found a way around this difficulty. There were up to four players allowed on each team, and from the next week onwards, on the two Mondays of the year when Miyuki was in the pub, they would put an extra pound in the pot, and though their team name remained Hughes Puw Hughes, their list of members would read, as it did tonight:

1. HUGHES
2. PUW
3. HUGHES
4. JAPANESE GIRL

And so even though Septic Barry had noticed the movement on the other side of the pub, he knew there was nothing that he or any of the Children from Previous

Relationships could do about it apart from pull faces and curse under their breath. Her name was on the sheet, and that was that.

Miyuki was tired after her early start and her train and bus journeys. She could feel her eyelids starting to droop, and after the final question she decided not to hang around as the quizmaster tallied the scores. She already knew they weren't going to win the round of beer. On quiet nights tall Mr Hughes, short Mr Hughes and Mr Puw had been known to lurch to victory, but The Four Walkers from Usk had looked like retired schoolteachers, and when the scores had been read out after the fifth and penultimate round they were ahead by an unassailable distance. Septic Barry and the Children from Previous Relationships were already quite happily resigned to their defeat. The only time they had ever won was when flu had been sweeping through the village, and having somehow escaped the bug they were the only team to turn up. They had claimed their round of beer after scoring just twelve points out of a possible sixty.

Miyuki took her empty glass back to the bar and said goodnight to her team mates.

'Goodnight,' she said to tall Mr Hughes.

'Be good,' he said.

'You too. Goodnight,' she said to short Mr Hughes.

'Don't do anything I wouldn't do,' he said.

'I won't.' She wondered what she would be restricting

herself to if she truly meant it. She thought she might try it as an experiment one day.

'Goodnight,' she said to Mr Puw.

'Sweet dreams, Thunderthighs,' he said, raising his pipe before relighting it yet again.

She raised her hand to Mr Edwards, who nodded back, and to Septic Barry, who lifted his glass in return and said, 'Good to have you back.'

She smiled. It was good to be back.

Back at the cottage she put some kindling and two small logs into the wood stove. There was still a patch of embers from the fire she had lit when she arrived, and the wood was dry so it caught after just a few blows. Some cold ash blew up her nose, and she sneezed her first sneeze since leaving home that morning. She turned on Radio 3 and half-listened to a contemporary Romanian madrigal as she licked her fingers and plucked out her contact lenses. She dropped the little pieces of plastic on to the top of the stove, and watched as they hissed and danced the mambo. Before long they ran out of steam and lay still, a pair of glossy lentils. She turned off the radio, and as she read the last few pages of her book she found that the murderer wasn't who she had thought it was after all, and felt a bit stupid for having missed so many clues and fallen for so many red herrings. She brushed her teeth and climbed into bed, where she wriggled out of her clothes and dropped them onto the floor, and pulled on the long johns and long-sleeved T-shirt that would be her pyjamas for the coming fortnight.

Before the four walkers from Usk had reached the bottoms of their winners' pints, Miyuki was asleep. With nothing in particular to worry about, and nobody to elbow her in the back, she slept very well.

TUESDAY

Miyuki started the first full day of her stay as she always did. She woke up a bit later than she would have done at home, and spent a while rubbing her eyes and staring at the ceiling. When she felt she had done enough of that she wriggled out of her makeshift pyjamas and put on several layers of clothes while still under the duvet. She got to her feet and shuffled through to the miniature kitchen in the corner of the sitting room, where she put three thick slices of white bread under the grill and fried half a bag of onion rings in margarine. She laced the food with ketchup and brown sauce, and ate it fast, washing it down with a cup of strong black tea. After brushing her teeth and sluicing the sleep out of her eyes, she filled up her water bottle, strapped on her boots, stepped outside and walked out of the village along the coast path.

There was a light morning haze, but not enough to mask the blue of the sky. After a while she left the main trail to follow a faint track of flattened grass that had been left by the feet of various animals and the boots of other walkers. They led to the end of a promontory, where she stood a couple of steps back from a sheer drop. A breeze was blowing in from the sea, and she faced it without blinking as tears ran sideways to her earlobes. Taking deep breaths, she closed her eyes and listened for a long time to the waves as they pummelled the cliff. When she opened them again the haze was almost gone, and the day was turning out just the way she had spent eleven and a half months hoping it would.

Back on the main path she met a small and amiable gang of wild ponies, and was glad that she had remembered to put a carrot in her pocket. She snapped it into chunks, and made sure each of them got a piece before she carried on along the cliff tops. She walked for miles without seeing another soul, and it felt as if the coastline belonged to her alone. In the summer the place would be crawling with surfers, climbers and sunbathers, the narrow roads jammed with cars and the path submerged beneath lines of walkers. She was glad she was able to go there in the middle of winter.

Feeling a thirst coming on she checked her map and walked inland to a pub in a small village. She got there to find she was their first customer of the day. They hadn't lit the fire, and the air carried a faint smell of toilet cleaner.

As with most of the pubs in the area, she had been in a few times before. She exchanged a smile of recognition with the woman behind the bar, and abandoning the vague rule she had about never drinking before noon, ordered a pint of OSB.

When it appeared on the bar the glass was full to the brim. Her hands were small, and she had to use both of them to get a steady grip and bring it to her lips without spilling any. As she slowly lifted it up and dipped her head to meet it she was reminded, as she often was in this situation, of footage she had seen of Japanese tea ceremonies. The movement came so naturally to her that she couldn't help feeling as though the muscles it used must have contained the ghost of an ancient memory. She pictured herself in a kimono, then felt a familiar sharp stab of embarrassment because all she was doing was drinking the top of a pint of beer, and that had nothing at all to do with being half Japanese. Once she had swallowed the first mouthful and there was a clear half-inch at the top, she was able to use one hand to lift the glass without the risk of spilling anything. She went outside and sat at a picnic table, where she read the first chapter of that day's book, and had some difficulty eating a bag of dry roasted peanuts with gloves on.

After her second pint she left the pub and walked back to the coast path and towards the cottage. She found herself wondering whether anything dramatic would happen over the coming fortnight. Maybe she would find a barrel of whisky that had washed ashore, or an abandoned baby by the roadside. She gave these scenarios some thought before deciding that if it came to it she would raise the child and drink the whisky, but she hoped she wouldn't have to. She didn't come here in search of adventure.

Her boots had picked up quite a lot of mud, so she left them on the front step. Once she was inside she got the fire going and boiled the kettle for some instant rice, which she ate straight from the plastic pot. She left the stove's door open, and read some more of her book as the flames warmed her feet.

Just before it started getting dark she put her boots and coat on, and walked back out to the coast path, in the direction of one of her favourite sights.

A short way from the village she walked across the grass to the edge of the cliff. Below her lay a small cove. There were plenty of small coves along this stretch of the coast, and a lot of them were surrounded by cliffs so steep that they couldn't be reached from the land. Miyuki would stand as close as she dared to the edge, peering down at the rocks and sand below, and wishing she could float down and explore. This cove was different, though. The cliff around it was still quite steep, but it wasn't particularly high, and was just about navigable. A narrow path had been defined by adventurous visitors, and it was possible, with some care, to get down to sea level.

She didn't do that though, at least not yet. Instead she stood where she was, taking deep breaths of the evening breeze, and waiting.

The air smelled and tasted different from earlier, more seaweedy. She supposed that had something to do with the tides. The sky was a clear, pale blue and the waves were low. Just as the sun was a thumb's width from the horizon, she felt a shiver as she saw what she had come to see. She had stumbled upon this sight on her first visit to the village, and had found herself returning to it year after year. Whenever the sky was clear at the end of the day, one of the rocks in the cove below looked, for just a few minutes, as if it had turned to gold.

She stood mesmerised for a while, but it wasn't long before she snapped into life and made her way down the steep, narrow path to see it up close. The top of the beach was made up of boulders, and she had to watch her feet as she crossed it, but the lower part was a mixture of sand and pebbles, with rocks poking out of it here and there. Grey rocks, of all shapes and sizes, and one big block of gold.

Miyuki had always reluctantly accepted that the rock only appeared to be gold from the cliff top because of a coincidence of light and angles and other things she didn't really understand, but a part of her still hoped that one day she would climb down to find it looking just as golden as it had from above. It was an almost perfect square about five feet across, and even though it lay above the seaweed and driftwood of the tide line, it seemed to have been smoothed by the waves. The setting sun still reflected brightly from it, but close to it was unmistakably grey, just like all the others. If anybody had been standing on the beach beside her they wouldn't have singled it out as being at all special, or even different.

As the reflected sunlight shone in her eyes, the unmistakable sensation of an emerging sneeze consumed her head, from the nose inwards. There was no need for her to stifle it, and when it came it was powerful and satisfying. It was swiftly followed by a more pedestrian explosion that brought her total for the trip to three.

She reached into her pocket for a tissue and blew her nose, then stepped up on to the rock and sat cross-legged as the sea swallowed the tip of the sun.

The dusk was long and slow, and there was still enough light for her to see her way to the top of the cliff and back to the cottage.

The evening meal she made for herself was so rudimentary that she felt slightly ashamed eating it. When she was finished she dumped the plate and pan in the washing-up bowl, put on her coat and went to The Anchor, where she sat in her usual place, underneath the stuffed pike.

She was fond of the pike. There was a small metal plaque on the wooden frame of its glass case, saying that it had weighed twenty-eight pounds and eight ounces on the day it had been dragged out of Llangorse Lake by a previous landlord of the pub. Her first moment of empathy with it had come when she noticed that the date on the plaque was the same month and year as her own birth. There was no precise day recorded so she couldn't tell if the fish was Aquarius or Pisces, but either way it was close enough for it to count as a contemporary. Moments later she felt stupid as it struck her that a pike of that size must have been swimming around for years, and rather than marking its birthday the plaque recorded the time of its death. By that point, though, a bond had been forged. To get around this difficulty, Miyuki counted the start of its new life as a pub ornament as a birthday, and she felt this still gave them some kind of affinity. Sometimes, as she returned to her table with a fresh pint, she exchanged what she was sure were knowing glances with the dead fish's glassy eye. A couple of years earlier, as thirty loomed for both of them, she had felt as though the fish on the wall knew exactly what she was going through, and that it too must have been asking itself questions about where it was heading, and wondering whether it was time for a new coat of varnish and a change of its slightly shabby diorama.

Over at the bar it was short Mr Hughes's round. 'Three of the usual,' he said to Septic Barry's girlfriend, who was working that night. She pulled him three pints of Brains.

Nothing had been said, but there was a palpable sense of surprise that she had lasted this far into January and was showing no sign of moving on. Usually by this time of year Septic Barry's girlfriends had gone back to wherever they had come from, or if they hadn't they would be looking sheepish as they plotted their imminent departure. This year, though, his girl was still around, and she seemed to be making herself more and more comfortable every day.

When he was sixteen, Septic Barry had argued with his parents over a jar of lemon curd. They had accused him of helping himself to spoonfuls of it – a charge he vigorously denied, his rage turning him alternately brilliant white and vivid purple. He had, of course, helped himself to the lemon curd, but that didn't stop him from being outraged by the accusation. He poured the contents of his savings jar into a sock, packed his rucksack, left the house and pitched his tent in a campsite a mile outside the village. He remained incandescent with fury as he lay in his sleeping bag and stared at the fluttering canvas walls of his new home. He worked out that for as long as he held on to his paper round he would be able to get by, and he pictured himself as an old man, still living in the tent and getting on his bicycle every morning to deliver papers.

The lemon curd incident soon blew over, and Septic Barry, or just Barry as he was then, often went back to his former home for meals that consisted of more than just cheap baked beans eaten straight from the tin, but he didn't see any reason why he should move back in, not now he had struck out on his own.

As he was busy failing his O-levels, the campsite started filling up. Families arrived, and old people, and cyclists, walkers, birdwatchers and foreigners, but most of the people who shared his field were invisible to Barry. He only had eyes for one kind of visitor.

He only had eyes for girls.

It hadn't taken him long to master the art of romance. He would see a girl he liked the look of, then stroll over and chat to her about whatever he thought she might be interested in chatting about before casually inviting her to drink some tepid snakebite or sweet white wine with him. Sitting beside her on a beach, a tree stump or a farm gate, he would impress her with tales of his unconventional lifestyle, and slowly inch closer as he recited streams of jokes he had memorised from the back pages of the *Daily Star*. If she seemed comfortable to have him so near that she could feel his breath on her face, he would lean into her, and if she leant back into him he would put one arm around her, and then the other, and he would kiss her, and because she was on holiday she would kiss him back in spite of his hair, and they would become a writhing, slurping Rodin.

It wasn't long before Barry's old sock was empty. His paper round wasn't making him enough money to pay for his spot in the campsite *and* for all the baked beans, snakebite and sweet white wine, and he realised he was going to have to get a better paid job if he was to maintain his high standards as a *bon viveur*. One afternoon he fell into conversation with the man who came to sort out the removal of the site's human waste. While they stood watching the thick black pipe quiver as the sludgy excreta of hundreds of holidaymakers shot up it and into a tanker, Barry happened to mention that he was on the lookout for odds and ends of work, and so it was that at the age of sixteen and a half he handed in his notice to the newsagent and moved into a whole new industry.

His new name followed later the same day.

When summer ended and the campsite closed to the public, the owner grudgingly let him move into one of the static caravans for the same money he had been paying to pitch his tent. He spent his winter thinking about the girls he had met, memorising even more jokes from the back pages of the *Daily Star*, and drifting deeper and deeper into human waste.

After a few years he had learned more or less all he would ever need to know about the installation and maintenance of septic tanks, and when his boss retired Septic Barry took over the business. He got himself a bank loan to buy all the equipment, and had the tanker and the van repainted. SEPTIC BARRY, they said on the side, in big blue letters, A GREAT SUCK-CESS SINCE 1994. This boast wasn't particularly impressive in 1994 itself, but as the years went by and he stayed in business, people quietly admired him.

Even though he worked five and a half days a week he still found time for the finer things in life, and whenever the site reopened in the spring he could be found slobbering all over these finer things, sliding his hands under their tops and all over their soft and wonderful bodies. They stayed for a few days, sometimes a week, and in rare cases even a fortnight, and when each girl was gone, the moment her bus had rounded the bend, he started looking for the next one to talk to, to drink with and, if the moment arose, to lean into. This went on right up until the terrible day the CLOSED sign went up for the winter and he found himself alone again.

At the end of one season Septic Barry decided to get in touch with some of his favourite girls from the summer to see if any of them would be interested in spending the cold months with him. Sitting at the foldaway table in the caravan he rummaged through a pile of torn-up cigarette packets with names and addresses scrawled on them in lip liner, and tried to remember which details belonged to which girl. Once he had a reasonable idea of who it was he was writing to, he would send a short *how-are-you?* note. Most of them didn't reply, but a few remembered their time with him fondly, and letters began to arrive.

The first was from a girl who had decided that she wasn't cut out for college and was wondering what to do with her life, the next was from one who had lost her job and was tired of being nagged by her mother to go out and find another one, and the one after that was from a girl who was just tired of her boyfriend and wanting a change. Before long he had quite a pile, and when he had decided which of his respondents he liked the best, he invited her to live with him in his caravan. She declined, but undeterred he moved on to the next one, who also declined. The next one, though, agreed to give it a try, and a couple of weeks later he drove the tanker down to the bus stop to await her arrival.

Year after year he kept to this strategy, and he found that he could always rely on one of the girls to accept his invitation. Even now, when it was getting harder and harder for him to get away with describing himself as *mid-thirties*, he would make his annual trip to the bus stop and find himself carrying an overstuffed holdall to the tanker for the short journey back to his caravan. Invariably short of money, his visitors would end up taking on shifts at the pub.

One year, the girl who came to stay was the most extraordinarily beautiful creature who had ever been seen in the village. She was incredible. So many people, on walking into the pub and seeing her for the first time, would involuntarily exclaim, *Jesus Christ!* that she assumed this was a customary local greeting, and without thinking she started to use it herself. 'Jesus Christ!' she would cheerfully say, as people came in from the cold, 'What can I get you?'

Mr Puw was among the first to encounter her. Before he had a chance to stop himself, he said, 'Jesus Christ!'

'Jesus Christ!' replied the girl. 'What'll it be?'

The vision had thrown him into such a state of confusion that he completely forgot what his usual drink was. Knowing he had to say something, and his eyes being on a bottle of Vermouth, anywhere being preferable to the glowing face and compact body before him, he panicked and ordered a sweet Martini and pineapple. When asked if he wanted ice and a slice, he just nodded. Septic Barry's girlfriend even added a cherry and a parasol. 'There you go,' she said, as she gently placed it on the bar. He knew he was supposed to thank her for the drink, but he couldn't move his mouth, and wasn't able to say a word as he handed over his five-pound note and took his change.

Shortly afterwards, short Mr Hughes arrived. Like Mr Puw he wasn't at all inclined towards blasphemy, but he took one look at the new barmaid, and spluttered, 'Jesus Christ!'

'Jesus Christ! What are you having?'

He found himself ordering a spritzer with a dash of lime and a chunk of cucumber, and when a similarly disorientated tall Mr Hughes arrived, he too took his saviour's name in vain before choosing a peach schnapps and grapefruit with a sprig of mint. They stood in virtual silence, trying as hard as they could to carry on as if nothing was amiss. Unable to say much besides *Same again*, they stayed on their original drinks all night. Occasionally one of them would half-heartedly mutter something about how a change is as good as a rest, the cliché thudding awkwardly around the quiet bar.

When Mr Edwards came down to see how his new barmaid was getting on, he spotted them with their colourful drinks, shook his head and said, 'Holy mackerel.'

The next evening, knowing to brace themselves for the onslaught of golden hair, dazzling teeth and sparkling blue eyes, they moved back on to their usual pints of Brains, recovered their manners and never spoke of their transgressions again. Her presence was inescapable, though. Her nights off should have been an oasis, but she couldn't be trusted not to come in and sit with Septic Barry, so at any moment the door might open and she could walk in, smiling and saying things like, '*Hello,*' or, '*How are you?*'

On both sides of the pub the regular customers didn't think it was quite right to be in the presence of somebody like her without smelling good, so they rummaged in the backs of bathroom cabinets for long-abandoned bottles of Old Spice, Brut and Blue Stratos, unwelcome gifts that had at last found their place in the world.

Even Septic Barry had been shocked when he met her at the bus stop. 'I remember her being nice enough,' he confided to the dazed-looking Children from Previous Relationships, as they watched her at work behind the bar, 'but I don't remember her being like *that*.'

She hadn't even been close to the top of his list for winter invitations, and he had made his way through quite a pile of torn-up cigarette packets before reaching her address and thinking he might as well give her a try. He had remembered a pleasant, shy girl who had been holidaying with a small group of friends as they waited for their A-level results. Her voice had been gentle, her body slim and her lips soft, but that could have been said of any number of girls he had met that summer. Something had happened to her since her holiday. She only ever wore simple clothes, usually just jeans and a pullover, and a little light make-up, but she was luminous.

As hard as people tried, they couldn't stop themselves from imagining her at it with Septic Barry. She kept appearing in their minds' eyes, arching her back and sighing as he pawed her exemplary breasts, blonde waves crashing over her naked shoulders. And this was Septic Barry, who cut his own hair and only ever wore budget jeans from camping shops and promotional sweatshirts from manufacturers and distributors of sewage disposal supplies.

Miyuki, who liked to think that she was immune to blondes, fell victim to her as well. 'Jesus Christ!' she said, on walking into the pub and seeing this half-girl-half-Roman-candle for the first time.

'Jesus Christ!' smiled the girl. 'What can I get you?'

A flustered Miyuki just about managed to get the word *snowball* out. There was no *please*, no *thank you*, just *snowball.* It was a drink in which she had never had any interest.

As the evening crept on, Miyuki recovered her manners and maybe even over-compensated for her earlier lapse, but she couldn't do anything except sit there nursing glasses of the sickly, custard-flavour drink, and pretending to read her book while casting furtive glances at the girl behind the bar, at her faultless skin and the shape of her breasts, at the way she wrinkled her nose when she smiled, and at her back view as she strolled around collecting glasses and emptying ashtrays.

She could hardly sleep for thinking about her, and the next day it took all the strength she had to stop herself from going to the shop to see if they had any bottles of Blue Stratos. As she got ready to go to the pub she rummaged through her suitcase and looked at the clothes she had brought with her. They all seemed so shabby.

When she got there, having carefully rehearsed her order for a pint of bitter, the girl had gone. Miyuki sat silently, burning with shame. She was supposed to be in love already, not having violent crushes on teenage barmaids. She knew she would never have done anything about it, even if she had been able to, but she still felt wretched. Convinced the stuffed pike was giving her disapproving looks, and knowing he was in the right, she drank far too many pints of Brains that night.

None of Septic Barry's winter girlfriends had anticipated the dusting in the pub, the stock rotation, the interminable conversations with lunchtime drinkers, or the smell of metal polish that was so hard to scrub from their fingers. As the December wind rattled the caravan at night and crept in through gaps in the windows and doors, they all concluded that whatever it was that they had been running away from wasn't as bad as all that. Their New Year's resolution would invariably be to leave, to clear out on the first of January itself or not long afterwards. When the blonde girl had gone, leaving a note in the caravan (*Oh Septic Barry, I'm so sorry!*), everybody breathed a sigh of relief because at last things could begin to go back to normal.

Although on balance he had wished she was still around, even Septic Barry felt a weight lift from his shoulders. He had never quite been able to relax enough to truly enjoy the time he spent with her.

A few weeks after her departure the scent of gentleman's cologne that had insinuated itself into the carpet and the wallpaper began to subside, and Septic Barry's ex faded with it into the past. Everybody assumed she had left to become a pop star, a toothpaste model or an actress, and for months afterwards every time they turned on the television they half-expected to see her smiling out at them. It never crossed their minds that she might have gone back to live with her mum and dad in Cwmbrân, found a job in a shop and got back together with her ex-boyfriend, who forgave her the moment she asked him to, even after everything he had said.

This January, though, Septic Barry's girlfriend was still around over a week into the month. She smiled at him as she pulled his beer, she patted his hand across the bar in quiet moments and when they left to go back to the caravan at closing time she took his arm. She was friendly and normal, and nice-looking without being so nice-looking it was weird, and although nobody liked to ask, it was clear that she was a more suitable age than his usual visitors. None of them could quite work out what somebody who seemed so well balanced was doing with Septic Barry, but then none of them could ever quite work out what *anybody* was doing with Septic Barry. It always seemed to make perfect sense when they left, but that just didn't seem to be happening this year.

Miyuki read the final page of her scruffy paperback, closed it and put it on the table. She made sure she read a book every day for the two weeks of her stay. She loaded her bag with slim- to medium-sized volumes, and by the time she got home she would have the satisfaction of knowing she had read an average of more than one book a month for the whole year, before it was even February. She sat quietly with her pint and her salt and vinegar flavour peanuts, without thinking of a great deal.

During her stays in the village most days would pass with her hardly saying a thing. One night as she lay in bed waiting to fall asleep, she realised that all she had said for that entire day was, 'Please,' when Mr Edwards pointed at the Brains pump, and, 'Thanks,' when he handed over her change. She had drunk four pints that night, so that totalled eight words. Normally she would be a bit more talkative as she thanked strangers for holding open shop doors, greeted passing walkers with a word about the weather, or simply drank more beer, but even though the eight-word day was exceptional the count rarely exceeded a few dozen. Calls home weren't allowed, and she was content to sit quietly and mind her own business. She wasn't interested in getting involved in the lives of the people around her – they were just part of the scenery, and it was only when Septic Barry sat beside her that she had anything that came close to being a proper conversation.

Over at the bar it was Mr Puw's round. 'More of the same, Thunderthighs,' he said, and Septic Barry's girlfriend pulled three pints of Brains.

Mr Puw called all women *Thunderthighs*. Miyuki had perfectly presentable thighs, as did Septic Barry's girlfriend. If anything all four of their thighs were above average in quality. Once they realised that there was nothing personal in what he said, that he was under the impression that it was a charmingly playful term of endearment, they became used to it and didn't really notice when it happened. Every once in a while though, Mr Puw would meet a woman who didn't have the slimmest legs. Having been greeted by him in his customary way she wouldn't be able to stop thinking about it, and would sleep badly that night, and not feel any better in the morning.

Miyuki dabbed her right forefinger around the peanut packet, collecting the leftover flakes of salt and monosodium glutamate and pressing them on to her tongue.

She half-listened for a while to some advice about the best way of responding to an alligator attack, then nodded her goodbyes to tall Mr Hughes, short Mr Hughes and Mr Puw as they drained their glasses, put on their coats and caps and headed home. She stuffed the empty peanut packet into her glass, which she took back to the bar.

She had meant to follow the others out into the night, but she reminded herself that she was on holiday so instead she ordered another pint. As she took her change she asked Mr Edwards if he had a Yellow Pages she could look at. He rummaged around beneath the counter for a while, and eventually resurfaced clutching a splayed copy that was covered in smears of spilled beer and cola syrup. It looked as though it had been down there for decades, and she was surprised to see it was the current edition.

Thanking him, she took it over to her table and turned to the *P* section.

Miyuki had rented the same single-storey, one-bedroomed terraced cottage for as long as she had been going to the village, but she still hadn't quite got the hang of its temperamental lock. The Brains was only 3.7%, but it all added up, and after her extra pint she fumbled in the dark for a full minute before getting back inside. She put a log on the fire, and sat watching the flames through the glass.

Her contact lenses were starting to itch. She didn't really enjoy touching her eyeballs, and back at home she could rarely be bothered with contact lenses. Usually she wore a pair of National Health glasses, which she sponged down with white spirit at the end of the day before running them under the tap, drying them on her sleeve and putting them straight back on her face, but on these trips glasses were an annoyance. They were always threatening to blow away in the wind, and even on relatively still days the sea could be magnificently rough, thundering into the rocks, and up in grand white plumes as it tends to in oil paintings. Even if the salty spray was so fine she could neither see nor feel it, it would keep misting up the lenses.

She pinched the little bits of plastic out of her eyes, dropped them on to the top of the stove and watched as they hissed and danced the tango, then curled and uncurled like maggots before giving up and lying still. She sat down in the armchair, and as she watched the flames her eyelids began to fall.

Some time later she woke up. She shuffled to the sink, where she drank a pint of water and refilled the glass for her bedside. She set her alarm to make sure she would be on time for her planned expedition, then quickly changed into her night clothes, switched off the light and flopped belly-down onto the double bed, where she spread out like the cross of St Andrew because she could.

WEDNESDAY

Between half past eight and a quarter to nine in the morning, a pale Miyuki Woodward could be seen standing at the bus stop by the café. She was reading a book, repeatedly yawning into the back of her hand, and holding an empty red canvas shopping bag that she had found in the cupboard under the sink. When the bus came she got on and paid the driver, and she wasn't seen again until after midday when the same bus, with the same driver, appeared from the opposite direction and pulled up at the stop across the road from the café. She got off, her book in one hand and a full red canvas shopping bag in the other.

A few people noticed her as she walked back to the cottage, but none of them thought to wonder what was in the bag. If they had done, they would never have guessed that it contained seven large cans of gold spray paint.

She went straight to the bedroom and put the cans in the bottom drawer, then boiled the kettle for a cup of tea and poured a packet of Worcester sauce flavour crisps into a bowl. As the tea was brewing she made a peanut butter sandwich with thick-sliced white bread. All she had eaten so far that day was a Galaxy bar and a packet of nice 'n' spicy Nik Naks, and she was ravenous. Before she put the jar away she couldn't resist scooping out a generous spoonful of peanut butter and eating it in one go. As soon as she swallowed it she felt it blocking her windpipe, and as she fought for breath she could only think of what an embarrassing way this would be to die. After thumping her chest and swigging some much-too-hot tea, her breathing slowly returned to normal.

Sitting at the small table by the window, she made a point of eating her sandwich in a more sedate, less life-threatening manner.

It was sunny outside and she didn't want to waste the rest of the daylight, so as soon as she finished eating she stepped back into her walking boots, put on her coat and went out, not along the coast path this time, but inland through the fields.

She walked down narrow roads, along farm tracks and footpaths, past isolated houses surrounded by ramshackle barns, around the edges of herds of cows and along the banks of ponds. The sky was blue, but in some places the ground was still sodden from the last rainfall, and her boots were soon so caked in mud that she couldn't see the laces. Looking down at her feet as they sank into suspiciously copper-coloured puddles of watery earth, she knew she should have worn her Wellingtons but she wasn't going to let it bother her.

As she was about to climb a stile somewhere around the middle of nowhere, her heart jumped when she saw a fox crouching in the long grass just a few yards away. It was looking towards the far side of the field, and she couldn't tell if it had sensed her. She stood still, just staring at it. It hardly moved, only the occasional twitch of its nose confirming that it wasn't a taxidermist's practical joke. Suddenly it turned its head to the left, then darted away in the opposite direction. She climbed the stile and stood on the top step, watching as it crashed through the grass and vanished under a distant hedge.

Moments later it became apparent what had startled it, as tall Mr Hughes came striding into view, holding a branch as a staff. She wished he wasn't there, that she could have been mesmerised by the fox for longer.

Tall Mr Hughes noticed her as she jumped down from the stile, and he lifted his stick as a greeting. She raised a hand in reply, and when he reached her he stopped.

'I remember when all this was fields,' he said, surveying the landscape around them.

'It still is fields, Mr Hughes,' said Miyuki.

'I know. And it was back then, too.'

Miyuki didn't know what to say, and she was relieved when he walked away.

'Watch out for alligators,' he called, as he climbed over the stile.

'I will,' she replied, but her voice didn't come out as loud as she had meant it to, and she wasn't sure if he had heard her.

She lost track of time, and was still some way from home when the sun started to go down. It became too dark for her to see where she was treading, and she rummaged in her bag for her pocket torch to find that the batteries were flat and she had left the spares in the cottage. The last mile was a miserable slog punctuated by squelches into unseen pools of liquid earth. As layer after layer of mud built up she felt her legs dragging behind her, as though she was wearing a pair of deep-sea diving boots. Water had managed to insinuate its way into her socks, which slowly made their way down her feet and sat, defeated, in soggy bunches around her toes.

She tried hard not to think about home, but it was no use. Grindl would be getting in from work around this time, making herself something nice to eat and lighting the fire ready for a night in with the television and a book. Or maybe she wouldn't be doing that. Maybe she had plans to meet up with friends and was sifting through the wardrobe, soaking in the bath or sitting in front of the mirror, expertly engaged in elaborate rituals that Miyuki could never quite get the hang of. She wished she could have been with Grindl, whatever she was doing, but instead she had chosen to trudge alone through the dark.

She sneezed. It was her sixth sneeze of the day, and the sixth time there had been nobody there to say *bless you*.

By the time she got back to the cottage her jeans were drenched to the knees. Telling herself that it would do her good to have a night off the beer, she decided not to go to the pub that evening, but to stay in by the fire and get to bed early.

She stoked up the wood stove and put her boots beside it to dry, then took off her jeans, ran the legs under the tap and scrubbed the denim with the washing-up brush. When they were more or less clean she hung them over the back of a chair to dry. After three cups of tea, a shower, a chapter of her book and a jacket potato topped with spaghetti hoops, she started to feel alive again.

She submerged the pan and plate in the washing-up bowl, and put on her dry jeans and the battered pair of trainers that she had packed for such situations. As if there had never been a change of plan, she walked down to the harbour and into The Anchor.

She sat at her usual table in the corner of the lounge

bar, underneath the stuffed pike, drinking pints of beer and reading her book. At around nine o'clock, Septic Barry left the Children from Previous Relationships and came over to see her. He sat down, and she folded the corner of the page she was on and put her book on the table.

'How are things with you?' she asked, knowing exactly what his reply would be.

'Oh, you know. Same old shit.'

She smiled. 'The old ones are the best.'

'Pint?'

'Rude to say no.' They always shared a few drinks over the course of the fortnight.

He went to the bar and ordered two pints of Brains from his girlfriend, who seemed to be perfectly comfortable about him buying drinks for another woman. Miyuki wondered whether Septic Barry had told her about the episode that had happened the year before. He probably hadn't. There was no particular reason for him to, and besides she wondered if he would even remember it.

Three days into Miyuki's last stay, Septic Barry's girlfriend of the time, a jovial Geography dropout with a big face and big features to match, had packed her bag and gone back to live with her mother and stepfather in Builth Wells. He didn't spend any time bemoaning his loss, and the moment she was out of the picture he turned his attentions to Miyuki.

That evening he had appeared beside her, and amid the usual chatter about the vicissitudes of human waste, and his analysis of various items that he had read in recent editions of the *Daily Star*, he made a point of letting her know that she was welcome to visit his caravan and listen to some of his new material. In itself this would have seemed like nothing more than a friendly invitation, but as he made the offer he crumpled his eyebrows, lowered the tone of his voice and appeared exaggeratedly casual, which told her that he would want to do quite a lot more than just play her his new material.

This shift in their drinking partnership took her by surprise, and her brain hadn't quite been running fast enough for her to work out the appropriate response to these unexpected overtures. She found herself agreeing to pay him a visit; all she could think to do was to leave the details as vague as possible in the hope of never having to go through with the ordeal.

'I'll definitely do that at some point,' she said.

Septic Barry was clearly very satisfied with the outcome of the conversation, and rejoined the Children from Previous Relationships with something of a swagger.

The following day he reappeared by Miyuki's side. She bought the drinks this time and they sat there once more, their Brains in front of them. This time the conversation didn't flow as freely as it usually did, and as they neared the bottoms of their glasses Septic Barry started to rummage in the right-hand pocket of his jeans. Miyuki assumed he was scratching himself, and discreetly looked away, but after a long struggle his hand emerged clutching a cassette.

'I've written a song for you,' he said.

'Oh. Thanks.' She took the cassette. It was warm, and seemed to be slightly damp as well.

He shrugged. 'It's just a little something I recorded at home. I hope you like it.'

'I'll give it a listen,' she said.

Septic Barry went back to the other side of the pub, and Miyuki looked at the tape. He had written on the label, in pencil, *I Don't Wanna Die (BHCYEMTGOL-WYIML?).*

Oh God, she thought.

Back at the cottage Miyuki braced herself, and pressed *play*. There was a loud click, then a steady wave of hiss mixed with the rumble of the motor of Septic Barry's tape recorder. He cleared his throat, and began. It was just him and his guitar, and the words were reasonably clear. She was able to work out from the chorus that *BHCYEMTGOLWYIML?* stood for *But How Can You Expect Me To Go On Livin' Without You In My Life?* She had to admit to herself that the song, just Septic Barry accompanied by his acoustic guitar, wasn't as bad as all that. It wasn't any good, but it wasn't terrible either. The tune was predictable but pleasant enough, even though it seemed to owe more than a little to Michael Jackson's

Earth Song – at one point during the ad-libbed wailing towards the end she was sure she could hear the words, *What about elephants?*

But it was the rest of the lyrics, the ones that weren't inquiring after elephants, that worried her the most. They told of how he could see the beauty within her but could see no beauty without her, and seemed to be a desperate plea for her to love him so that he might not perish as the direct result of a broken heart.

Septic Barry had faded the song out by carrying on singing and playing while moving away from his cassette recorder, through the caravan's front door, down the steps and over to the other side of the field. When only the hiss and the rumble remained she pressed the stop button, and wondered what on earth she should do.

With the addition of *I Don't Wanna Die (But How Can You Expect Me To Go On Livin' Without You In My Life?)*, Miyuki had five songs that had been presented to her by misguided suitors, and Septic Barry's was one of the more accomplished efforts. It wasn't quite up there with her favourite, *So Exotic*, a light calypso number that extolled her spellbindingly international genealogy, but it was streets ahead of the worst one, *Now My Secret Has Been Revealed*.

The first song that had been given to her, *Now My Secret Has Been Revealed* had landed on her doormat early in her first term at college. Written and performed by a pale boy called Christopher from one of her tutorial groups, the cassette was accompanied by a note in which he confessed to being too shy to speak to her, and asked her to simply play the song and realise how he felt. He begged her never to speak to him unless it was to tell him that she felt the same way.

She lay on her bed, put the tape into her Walkman and listened. He accompanied himself on a keyboard that seemed to be set to *Sounds of the Jungle* as he shrieked in the most horrendously impassioned way for seven and a half minutes. *Oh no*, she thought, when the song finally ended. *What do I do now?*

She had no idea. For the next three years, every time she saw him approaching she crossed the street or ducked down a corridor, or if it was too late to hide she simply looked at the ground as they passed, doing a bad job of pretending she hadn't noticed him. After giving her the cassette he had transferred to a different course so they never shared another class, but they graduated at the same time, and after their final exams he sent her another note, without a song this time, thanking her for her kindness towards him for the preceding three years. As a postscript he told her that his feelings hadn't changed, and that they never would.

She knew from her own miserable experience that romantic fixations eventually wither into embarrassing memories. She told herself that now Christopher was free from the constant threat of her appearing from around every corner of the campus he would, at last, be able to move his life forward.

A few years after college she made one of her occasional trips to Cardiff, and saw him coming out of a chip shop. He hadn't crossed her mind for a very long time, and without really thinking about it she smiled at him as she would an old friend.

'Hello Christopher,' she said.

As he looked back at her his mouth at first fell open, then turned into a smile. 'Does this mean . . . ?' he said. His eyes began to shine. 'Does this mean . . . ?'

Miyuki realised with horror what she had done.

'Oh God,' she said, as she felt the blood drain from her face. 'No, it doesn't mean that. I was . . . I was just saying hello. It's nice to see you, that's all. I'm so sorry.' She walked away as fast as she could without breaking into a run, but she couldn't stop herself from glancing over her shoulder to survey the wreckage.

He was frozen, staring at the empty space where she had stood. She hadn't passed that way since, and as far as she knew he was still there on the pavement, his eyes fixed on a small patch of air as he clutched a bag of cold chips.

The evening after Septic Barry had presented her with his cassette, she sat underneath the stuffed pike knowing she had to face up to her responsibility and extinguish every spark of hope as quickly as possible. She wondered whether she should tell him about Grindl, if that would make things better or worse. She wouldn't even have to go into specifics, she could just say something like, 'I'm with somebody.' It would sound stark and ugly, and though it would break the rule she had imposed on herself about not talking to anybody in the village about her home life, it would make her situation clear enough. She didn't want him to be another Christopher, wasting his life clinging to impossible dreams.

He arrived and sat, as always, with the Children from Previous Relationships. Then, after a couple of pints, he made his inexorable way towards her.

'So,' he said, putting their Brains on the table and sitting down beside her, 'what did you think?'

'It was . . . quite good,' she said.

Wildly encouraged by this faint praise, Septic Barry reiterated the invitation to his caravan. 'I've got a load of others I could play you,' his voice getting lower and his eyebrows becoming increasingly crumpled as the sentence progressed, ' . . . in person. Why don't you drink up and come back to my place?'

The time had come. She took a deep breath. 'I'm really sorry, Septic Barry,' she heard herself say, 'but I can't go with you to your caravan. I just can't. Not now, not ever.' There was a horrible silence between them. 'Never.'

Septic Barry felt the situation slipping beyond his control, but he had one more weapon in his armoury – a carefully turned and well rehearsed compliment, designed to charm his quarry into submission. 'You know,' he said, 'ever since I first saw you, I've always thought that you were full of Eastern promise.'

This didn't work quite as well as he had hoped. Instead of melting into his arms she just stared straight ahead, her mouth hanging open.

'Oh. Right,' he said. 'Never mind. You can't blame me for trying though, can you?' With that he returned to normal, telling a convoluted anecdote relating to the intricacies of the disposal of solid waste, and giving his opinion of some of the exchanges he had read on the problem page of that day's *Star*. Before long he said goodbye, and went back to sit with the Children from Previous Relationships.

Later that same evening he reappeared beside her and asked if she would mind returning the cassette to him. 'I forgot to do a copy,' he explained. She had it in her coat pocket, so she was able to hand it back to him there and then.

For the days that followed, Miyuki worried that Septic Barry was smarting from his rejection. He went about his business as if no advance had ever been made, let alone rebuffed, but this could have been his way of coping with a heart in tatters. Even though she knew she had done the right thing, she still couldn't shake off a feeling of guilt.

On the last night of her trip she saw him talking to a

woman who had been in on her own the night before. She worked out from overheard snippets of conversation that this woman was a recent divorcee who was spending some time getting away from it all as she readied herself to rebuild her life. As she passed them on her way back from the toilet she heard Septic Barry telling the woman, in an exaggeratedly casual way, that he had written a song for her after seeing her the night before. His eyebrows were crumpled and his voice was low. A cassette changed hands, and as the woman thanked him and put it on the table, Miyuki glanced over and saw that written on the label, in pencil, was *I Don't Wanna Die (BHCYEMTGOL-WYIML?)*.

She half-smiled to herself, and left the pub without any goodbyes. Two weeks of nothing much at all interspersed with unwarranted concern about Septic Barry's feelings had exhausted her. She was glad to be going home in the morning, and to know she wouldn't be back at The Anchor for almost a year.

This year, though, there were no songs, no crumpled eyebrows and no invitations to his caravan. Instead he just chatted away quite happily about some forthcoming repairs on his van, and told her the story of a serial bigamist that he had read in that day's *Star*. Sometimes he smiled over at the barmaid, who smiled back.

'You really like her, don't you?' said Miyuki, after one of these wordless exchanges.

'She's OK, I suppose,' he said, but his face took on an almost dreamy quality that she had never seen before, and which quickly vanished like a bashful child in a room full of grown-ups. 'She says she sees potential in me. I don't know what she's talking about – I haven't got any of that.'

'Don't be so sure,' said Miyuki. 'You might just surprise yourself one of these days.'

'I hardly think so. What you see is what you get with me.'

'Exactly.'

He looked puzzled. 'Exactly what?'

'Like you said, what you see is what you get. And she sees potential, so there you are.'

'I hadn't thought of it like that. Maybe she's right, maybe I have got potential. Maybe I could . . .' He slipped into a reverie.

Miyuki thought it best not to intrude on his thoughts, and they sat in a comfortable silence. Septic Barry's girlfriend kept glancing over at the two of them, and smiling. She had a really nice smile. Grindl had a really nice smile too, and from this weak thread of connection Miyuki found herself gripped with unhappiness.

She was battered with visions of what her life would be like if something terrible was to happen. Melodramatic thoughts piled on top of one another as she pictured Grindl being thrown across the room by a fatal electric shock from the faulty wiring of the house she was working on, then Grindl lying on a hospital bed, desperately ill from an agonising and incurable disease, able only to communicate by twitching her left big toe. She imagined Grindl bound and gagged as kidnappers stubbed out cigarettes on her body, and then, and this was the worst of them all, Grindl going off with somebody else, somebody she would love more than she had ever thought possible.

Miyuki felt as though her skin was shrinking as she saw visions of herself growing old in an empty home, or hopelessly trying to start a new life with somebody who wasn't anywhere near as right for her as Grindl, somebody with an annoying voice and who wore really bad earrings.

This assault subsided after a few seconds, leaving just a faint sense of embarrassment at some of the overwrought dialogue that had appeared in these scenarios, but it had lasted long enough for her to wonder, just for a moment, whether her response to Septic Barry's attempted seduction would have been different if she had been alone in the world. Maybe she would have gone back to his caravan and listened to his songs, and maybe when he leaned in to her she wouldn't have edged away. She couldn't know for sure that she wouldn't have leaned back in to him, and clung to him for warmth, and kissed him back in spite of his hair.

Miyuki sneezed. She looked at Septic Barry, hoping for a *bless you* to set her tally back to zero, but he was still lost in his trance and didn't seem to have noticed. She knew there was no point in prompting him, because the blessing had to be spontaneous for it to count. She added it to her running total, which had already reached the twenties. She blew her nose and went back to her beer, hoping she wasn't going to end up getting a full-blown cold.

'You know what?' said Septic Barry, suddenly snapping back to full consciousness. 'She could be the one.' The words sounded alien as he said them. He had never thought for a moment that he might be looking for *the one*. Still in confessional mode, he carried on. 'Maybe you're right,' he said. 'Maybe it's time for me to start making some big decisions.'

'Really? What about?'

'Oh, you know. This and that.'

Miyuki didn't press him to carry on.

Septic Barry decided he'd had enough of their serious conversation. 'Here's one for you,' he said, and he launched into a joke about a nun, a sex-starved poltergeist and a bottle of altar wine. Miyuki didn't quite get it, but she wasn't going to ask him to explain so she laughed anyway.

'That's not the punchline,' he said.

'Oh.'

He carried on, then stopped, but she couldn't be sure that he had finally reached the end of the joke so she stayed quiet.

Septic Barry looked at her, and she looked at Septic Barry.

'That was the punchline,' he said.

'Oh, right. Very good. I'll remember that one.'

'You didn't get it, did you?'

'No, not really.'

Septic Barry went back to the start of the joke and explained it to her, line by line.

'Ah, I get it now,' she lied. 'Very funny.'

Satisfied at last, he went back to sit with the Children from Previous Relationships, and she slowly carried on with her pint and her book. She tuned in and out as tall Mr Hughes talked short Mr Hughes and Mr Puw through a series of facts about the breathing habits of the alligator. 'And that's partly what makes them so bloody dangerous,' he concluded with a chuckle, apparently blind to their absolute lack of interest. 'Oh yes, they're dangerous all right, those alligators.'

Short Mr Hughes turned down the corners of his mouth, and nodded very slightly. Mr Puw carried on staring straight ahead.

Miyuki finished her drink and her book, and wondered for a moment whether she should get herself another pint, but she decided not to. She needed to get to bed if she was going to be up and out long before sunrise. She put her glass on the bar, smiled a few goodnights and left the warmth of The Anchor behind her.

She picked some kindling and a medium-sized log from the wood basket, put them on the embers and blew the fire back to life. She watched the flames for a while, and moved the log around with the poker for no particular reason, then went over to the sink and ran her hands under the tap. Going back over to the stove, she pinched out her contact lenses and dropped them on to the hot metal. She watched as they hissed and danced a waltz, then curled up and wriggled before lying still, a pair of blackened, petrified tears.

She turned on the radio. A man from Bergen was ringing a giant bell that he had cast himself. She hoped that the small speaker wasn't doing justice to his life's work because it sounded awful, like somebody hitting an empty gas cylinder with a curtain rail. She was relieved when the clanging stopped after a minute or two. This was followed by a live recording of a group of Bulgarians creating a soundscape in a Parisian art gallery. It didn't seem to be much more than a drone, and she could feel her eyelids drooping.

She didn't want to fall asleep in the chair again, so she forced herself up, turned off the radio, brushed her teeth and got into bed, where she lay on the side she slept on at home. She thought back to her moments of despair in the pub, and laughed out loud. It was these blasts of quiet mortification that were the main reason for her going to the village year after year. They usually started happening on the Wednesday, and although they weren't much fun at the time, once they had passed she was pleased to reassure herself that she wasn't wasting her time. She looked at the photograph that stood propped against a mug on the bedside table, and stroked the part of the sheet where Grindl would have lain, had she been there.

'Goodnight Grindl,' she whispered. 'Sleep well.'

She set the alarm on her digital watch, and closed her eyes. She kept thinking about what she had planned for the morning, and it took her longer than usual to drop off, but when she did she slept deeply, her right arm resting on the empty side of the bed.

THURSDAY

Miyuki was able to do most of the work by the light of the moon and stars, and as the sky turned from black to the darkest blue she used her seventh and final can of gold paint to fill in any patches she had missed. The adrenaline that had propelled her along for the first few cans was wearing off, and she could feel her energy draining away. Her forefingers were numb from holding down nozzles, and her thighs had begun to wobble from all the crouching as she shuffled around the rock like a giant crab. The soles of her Wellington boots had become gilded, and she was wondering whether they would look any good if she was to spray them gold all over when she heard a man's voice.

'You're up early,' it said, muffled by the hiss of the can and the rumble of the waves.

Oh Jesus, she thought. Her lungs froze, and she pressed a hand to her thumping heart. She turned, and could make out a black shape just an arm's length away. He was so close that the mist from his breath was mixing with hers in the cold air. From her perch on the rock their faces were on the same level. She kept a finger on the nozzle so she would at least be able to spray him gold if he was to lunge for her.

The sheen of the paint had shrunk her pupils, and as her eyes adjusted back to the darkness of the early morning she was relieved to make out the face of tall Mr Hughes. She felt a bit stupid for having almost been knocked unconscious with fright by somebody as un-terrifying as tall Mr Hughes. Her shock turned to worry though, mainly because he had caught her painting the rock, but also because there was a very real possibility that he would start talking about alligators, and it was far too early to have to listen to that. After a while she was able to breathe again. Her mind cleared and she found she was able to talk.

'Hello, Mr Hughes,' she said.

Her heart was still thumping, but without the urgency of before, and she let her hand drop from her chest to her side. Beside the rock she was painting was another one, like the bigger rock in miniature. She stepped down on to it, and looked across the gold at tall Mr Hughes, who was studying her work. The rock was glowing brighter all the time as the morning moved in.

'So what's this all about then?' he asked.

Miyuki had wanted it to seem as if good fairies had come down overnight and sprinkled the rock with magic dust. She hadn't anticipated talking to anybody about it, and her mind was blank. She had no idea what to say.

'Not answering? I suppose not. You're inscrutable, aren't you? You . . .' He wasn't sure what the next word should be. 'You . . .' He knew he wasn't supposed to say *Nips*. 'You . . .' He wondered if *Japs* would be OK. Again he wasn't sure, so he played it safe. 'You lot.'

Miyuki accepted that her silence could conceivably be mistaken for an air of inscrutability, and she hurried to find some words, any words, so she could try to explain herself. At last some appeared. 'It's art,' she said.

'Modern art, is it?'

She nodded.

'So what does it mean, then?'

'What do you mean, *what does it mean?*?'

'I thought modern art was all supposed to *mean* something, not just be nice to look at.'

Miyuki wondered if he was right, and that it should *mean* something. The trouble was she hadn't thought very deeply about it; she had only done it because she thought it would give the occasional passer-by a welcome surprise. If it really was art, as she had declared it to be, then maybe tall Mr Hughes was right, and it ought to be seen to be making some kind of pertinent point about the world around them.

'It's a comment,' she said.

'A comment, is it? I see.'

For a moment she thought she had got away with it, but tall Mr Hughes wasn't satisfied with her answer, and wanted to know more. 'What's it a comment on?'

'On . . .' It was still very early in the morning, and her brain hadn't fully woken up. 'On . . .' Tall Mr Hughes was looking at her intently. 'It's a comment on . . .' She desperately wanted to bite into her bottom lip but she knew that would make it obvious that she was making things up as she went along. Again it felt as if her mind had been wiped clean, but at last a topic of suitable gravity leapt from the whiteness. 'On mining.'

'Not much of that going on these days.'

'Well, it's not so much about the coal mines,' she said, as her brain warmed up and inspiration struck. 'It's more about gold mines.'

'What, like that one over by Llanwrda?'

She had forgotten about the gold mines of Wales. 'No, I mean the ones in Africa.'

'What about them?'

'About how horrible they are to work down. Things like that.'

'Oh, right. I suppose they must be, if you think about it. Hot, I'd have thought, what with them being in Africa.'

'And . . .' Her thoughts were racing now, and something else struck her as she looked at the shape of the rock from the corner where she stood, 'it's about diamond mines, too. It's a comment on how horrible it is for the people who work down those, as well.'

'Dwarves, mainly.'

'I wouldn't know about that.'

'Coloured people, then.'

'I'm not sure. I suppose so, though.'

'But mainly dwarves, like in that film. They fit down the holes better, see.' He fixed his eyes on the horizon, which had just become visible. Only the brightest stars were still shining.

The conversation had taken another difficult turn. Miyuki was already feeling ashamed for having spoken such rubbish about art and mines, and once again she had no idea what to say. She knew this wasn't the first time tall Mr Hughes's thoughts had turned to the subject of dwarves.

Tall Mr Hughes was an occasional contributor to the letters page of one of the local papers, and a few years earlier he had written in appealing to the people of the area to be compassionate towards those considerably shorter than themselves. His letter was selected for publication, and as people read it they were unable to argue with its key sentiments – they agreed that dwarves were people too, people with feelings, and as they were implored to not mock or ridicule *these terrific little characters*, they reflected for a moment and found themselves quite satisfied that they hadn't been harbouring any particular intention of being mean towards the very short. They would have forgotten the letter completely if it hadn't been for something not quite right about it.

The headline, which had been suggested by tall Mr Hughes in a postscript, begged its readers to REACH OUT TO DWALVES, and closer inspection revealed that he had misspelled the plural of *dwarf* all the way through. It began, *We have all chuckled at the antics of dwalves in the annual Christmas pantomime*, and went on to praise *these brave little chaps the dwalves*, and concluded by suggesting to his readers that the next time they were to find themselves peering over a garden fence they should *take a moment to reflect upon those whose very height precludes them from so doing, namely our diminutive friends the dwalves.* The editor, a notoriously shambolic brandy drinker, didn't seem to have noticed the mistake, and printed the letter exactly as it was submitted.

Nobody gave the piece much further thought until a few months later, when tall Mr Hughes was seen striding proudly through the village while beside him, almost running to keep up, was a young man who couldn't have been as much as four feet tall.

There were sightings of the pair of them all over the place – reading discarded newspaper supplements in the café, browsing in the mobile library, flagging down the ice cream van, and waiting under an umbrella for the bus to Haverfordwest. Nobody knew who the newcomer was, where he had come from, or what he was doing hanging around with tall Mr Hughes. In the evenings they turned up at The Anchor, but instead of standing with short Mr Hughes and Mr Puw, tall Mr Hughes would order a pint for himself and a half for his *little friend*, and they sat at the table in the corner of the public bar, spending the rest of the evening deep in hushed conversation.

This carried on for five days, until the moment the quiet of the bar was shattered by the bang of a half-pint glass on the table, and a shout of, 'Stop treating me like a fucking dwarf.' With all eyes on him, tall Mr Hughes's companion bolted towards the door and out onto the street, a flustered tall Mr Hughes in his wake.

That was the last time the visitor was seen in the village. Rumours went around about the nature of their friendship, and about what had happened to tall Mr Hughes's guest. The more innocent speculations had him as a visiting nephew, but others were more salacious. One story went that after a lovers' tiff tall Mr Hughes had slit his friend's throat, cut him into pieces and fed the little briquettes of dwarf into his wood stove.

After a couple of days of lying low, tall Mr Hughes was back in the pub, drinking with short Mr Hughes and Mr Puw, and delivering an extended monologue on the subject of submarines as if nothing out of the ordinary had happened.

Miyuki had missed all this. It wasn't until she returned over a year later that the episode became news again, and she heard all about it from Septic Barry. It turned out that tall Mr Hughes's dwarf, far from having been used as solid fuel, was very much alive and had been seen by everybody in the starring role of a wildly popular series of television commercials for oven chips.

His fame had caused a lot of excitement. Phone calls were made to relatives in which it was revealed, as an aside, that somebody they knew was on television. Long-dormant friendships were rekindled with several-page letters in which, a few lines before the end, casual mention was made of the connection. *Have you seen those oven chips adverts?* they would ask, knowing that the recipient couldn't possibly have missed them. *I know the dwarf.* People who had seen tall Mr Hughes's former companion go by in the street called him *a good bloke*, or *a laugh*. A teenage boy who had never spoken to him, and who had made an elaborate but unexecuted plan to knock him to the ground, went around describing him to people as *my friend*. And through all this tall Mr Hughes would not be drawn to comment, and showed not a flicker of pride or dismay.

Miyuki felt guilty for raising the subject of dwarves, before realising that she hadn't raised the subject of dwarves, it was tall Mr Hughes who had brought them into the conversation. 'Anyway,' she said, trying to move things to safer ground by returning to his opening gambit, 'you're up early too, aren't you?'

'I'm always up early. And I expect you will be when you're as old as I am.'

'You're not old.' She guessed he was somewhere in his sixties, but it was hard to say for sure. The plain truth was that tall Mr Hughes was at least fairly old, and she felt embarrassed for having denied this.

'Well, perhaps that's a matter of opinion. Anyway, I got used to coming down here early in the morning. If the tide was out I would throw sticks for my dog down by there.' He indicated the damp, flat sand between them and the sea. 'He would run up and down for hours. I've not had a dog for a few years now, but I still come down here most mornings if the weather's not too bad.'

Miyuki remembered that in her first year or two in the village tall Mr Hughes always had a world-weary mongrel slumped at his feet as he drank in The Anchor.

'I miss him, you know,' he said.

She was sorry that he had lost his companion, but also relieved that they seemed to have left art and dwarves a safe distance behind. 'You could get another dog,' she said.

'No, it's too late now.'

'It's not too late.'

'Yes, it is.'

'Well, maybe you could get an alligator then. You could get up early in the morning and walk it along the beach.'

Tall Mr Hughes gave her a look that she couldn't quite unravel, and she regretted having raised the subject of alligators. He didn't pursue it though, and before he had a chance to change his mind she hurriedly moved the conversation back a couple of steps. 'All I'm saying,' she said, wishing she could think of a completely different subject to move on to, 'is that I see plenty of people older than you out walking dogs.'

'And all *I'm* saying is that I'm too old for anything like that. It wouldn't be fair on the dog. Or the alligator. Who knows how much longer I'd be able to look after it? I'm on my last legs, I am. It's downhill from here for tall Mr Hughes.'

'You know that's not true. I see you out and about all the time. You've got years left in you. I bet I'll come back here in twenty years and you'll still be walking the paths and propping up the bar with short Mr Hughes and Mr Puw.'

'Them? I doubt they'd want me there, those two.'

'What do you mean? They're your friends.'

'Are they, though? Sometimes I can't help thinking that if I dropped down dead they wouldn't even notice.'

'I think they would notice. They like you, I can tell. Of course they do. It's just, well, they're men, aren't they? You lot don't really like talking about things like that, do you?' As she said this she realised she didn't really like talking about things like that either.

They looked at the rock, which was glowing more and more golden as the morning moved in. She had hoped to be safely back indoors by this time, and she was itching to get off the beach.

Just as it seemed as though the silence between them

would never end, tall Mr Hughes said, 'Look at it.'

Even though she was already looking at it, Miyuki looked a bit harder, pushing her head forward so he would see she was making the effort.

'It looks like a pot of gold, doesn't it?' he said. 'A big square pot of molten gold.'

Miyuki told him that she could see that, and they fell back into silence. She was glad that he seemed to be appreciating her work.

'You know what?' said tall Mr Hughes. 'If that really was a big pot of molten gold, I might jump in it right now.'

'What?' Miyuki had heard him perfectly well, but still couldn't quite believe it.

'I said I might jump in if it was really molten gold.'

'I thought that's what you said. But why?'

'Why do you think? It would swallow me up. That would be that. It would all be over with.'

Miyuki hated it when people talked like that, and a throb behind her eyes told her she was close to losing her temper. 'Well, I suppose it would all be over with. But it isn't molten gold, it's a rock. You'd just be covered in bruises and scrapes, and besides even if it was a pot of molten gold I don't believe you would jump in.'

'How do you know?' he asked.

'Just look around you. You hardly need a pot of molten gold if that's the way your thoughts are heading. Look how high the cliffs are. Look how deep the sea is.'

Tall Mr Hughes didn't seem to react to this. 'Sometimes I lie on the grass and fall asleep, and hope that by the time I wake up I'll have been torn to pieces by vultures.'

Miyuki swallowed hard as this image appeared before her. This wasn't the kind of talk she expected from tall Mr Hughes.

'You don't get too many of them round here,' she said.

'Puffins, then.'

'Do they tear people to pieces? I don't think they do.' She could feel her hands shaking, and her voice becoming combative. 'And anyway, you don't get them at this time of year. Even if they do feast on human flesh you'll have to wait until spring.'

'Well, I suppose . . .' His sentence tailed off, and Miyuki didn't push him to complete it. She was desperate to get home before anybody else saw her, and she hoped that tall Mr Hughes would forget this whole incident.

'I'd better be off,' he said, at last. He was looking with heavy eyes at the golden rock. Miyuki's anger had subsided, and she didn't want to see him go away feeling so despondent. The whole point of her plan had been for it to cheer people up. Tall Mr Hughes had been the first person to see the result of it, and now he seemed to have plunged into a state of despair.

'Will I see you in the pub tonight, Mr Hughes?' she asked, taking care to speak gently.

'I'm not sure that you will, young . . . young . . . what did you say your name was?'

'Miyuki.'

'Did I ever know that?'

'No.' Nobody in The Anchor ever called her by her name. Over the years he had gradually picked up the names of the regular customers by overhearing conversations, but nobody had ever asked her what she was called. She couldn't even remember having given her name to Septic Barry, who she almost thought of as a friend. 'No,' she said. 'I don't think you did.'

'Anyway, young . . . what was it again?'

'Miyuki.'

'Miyuki. Japanese, is it?'

She nodded.

'I thought as much. Anyway, young Miyuki, I'll be heading on my way now.'

'You shouldn't feel so down,' she said. 'Do something fun. Do something you wouldn't normally do.' As she said this she recognised it as inane and clumsy advice. 'I'll see you in the pub tonight, Mr Hughes. I'll buy you a drink.'

With a brief wave of his arm he turned and walked towards the steep, narrow path that ran up towards the cliff top. Miyuki watched him as he climbed it with the agility of a boy, his feet finding every foothold to propel him onwards, and knowing every hazard to avoid.

Once he was out of sight she hopped back on to the big rock and quickly sprayed the remaining spots of grey that were peeking through the gold. It was the first time she had used spray paint for years, and she had forgotten just how much dust the cans spewed out along with the paint and the fumes. She knew she should have got herself a mask when she bought the paint, but she hadn't thought to, and she had to put up with breathing the noxious air unfiltered. She sneezed so often, just small, light blasts, that she had almost stopped noticing when it happened and she worried she would lose count. She didn't though. By the time the final can was empty her total tally for the trip was up to fifty-one, and it was only the morning of her first Thursday.

She put the empty cans in the bag, and put her coat on. She no longer needed the torch, so she left it in her pocket as she carefully made her way back to the cliff top.

Lights were on in the houses, and smoke from freshly lit fires was coming from a few chimneys, but she didn't see anybody on the way home and she didn't think anybody had seen her. She was relieved to kick off her boots on the doorstep and get inside. She put the bag of empty cans under the bed, then ran a full sink of hot soapy water and went back outside for the boots. She plunged them in, and after letting them soak for a while started scrubbing them with a scouring pad.

As well as forgetting to buy a mask, she hadn't thought to get a bottle of white spirit while she was at the paint shop. Back at home she would spend the final minutes of every working day washing brushes and getting paint off her hands. It should have been second nature for her to think about cleaning up, but it hadn't crossed her mind until now.

As she scrubbed the gold and grey from under her nails, it almost seemed like any other day back at home.

Escaping from a short and disastrous attempt to establish themselves in the city, Miyuki and Grindl had moved to a small town in the Valleys, not far from where Miyuki had grown up. They bought a run-down cottage at the end of a terrace at the top of the steepest street, and decided to set themselves up as interior decorators. This plan seemed to make sense because Miyuki had been on an art course so she knew a reasonable amount about paint, and Grindl had taken Maths A-level so she could work out all the finances. They practised on their own home, and found that there was a lot more to it than they had at first thought. Keeping their day jobs, they enrolled in evening classes in decorating and DIY, and practised again on their own place, and Miyuki's mother's house, and on friends' spare rooms and landings. When they had practised enough they took on jobs working for other people, and when they finally dared to go it alone and start on strangers' homes they were pleased to find that they weren't too bad at it, and that just about enough work came in for them to be able to get by.

Buying a small and ancient van, they called themselves GM Interiors. This stood for Grindl & Miyuki, although because of an unfortunately timed slew of news headlines that had appeared while they were dropping leaflets through letterboxes, everybody assumed it stood for Genetically Modified. Either way, they were known locally as *The Lesbians*. Whenever a conversation arose about home improvements, somebody would inevitably ask, 'Are you getting The Lesbians in?'

Their clients would be visibly disappointed if only one of The Lesbians was to turn up for a job. They would try not to let it show, but it was obvious that they felt a bit short-changed. It always happened to Miyuki when Grindl was spending her annual fortnight away, and she was sure it would be happening to Grindl right now. 'Where's your friend?' they would be asking, as if it didn't matter at all, and they were just making polite conversation.

After a long struggle she got her boots and her hands free of gold, then loaded the washing machine with the clothes she had been wearing and the contents of the plastic bag she had been using as a laundry basket. She showered, climbed into some cleanish clothes, made a mug of tea and opened a family pack of bourbon biscuits. She hadn't eaten anything all morning, and her belly was crying out for food. She dunked a biscuit in the tea and ate it, then did the same with another, and another. She slipped into a trance, and by the time she regained full consciousness they were all gone, twenty-four biscuits in one go. It was something she hadn't thought possible, and along with a feeling of mild nausea came a real sense of achievement. Exhausted, and not nearly as ashamed as she supposed she ought to have been, she went to the sink and poured away the inch of gritty liquid that was left at the bottom of the mug.

She felt as if she was sleepwalking as she went through to the bedroom and flopped on to the bed. Seconds later she was asleep.

When she woke up she hung her damp laundry over radiators and over the backs of chairs, and after a brown sauce sandwich and a pair of mini Swiss rolls she strapped on her walking boots and headed along the road that ran out of the village and towards the long beach. She walked across the big and almost empty car park, sat at a picnic bench and opened her book, a short biography of an actress she had once adored but who these days only crossed her mind every once in a while. She had felt this waning of devotion in many aspects of her life, and she held Grindl responsible. Grindl took up so much of her heart and mind that her old obsessions and preoccupations had been starved of oxygen, and had withered into insignificance. Grindl had experienced similar changes, and had blamed Miyuki for the fading of all kinds of passions. She had been particularly annoyed when she realised that her favourite country songs no longer stabbed white hot skewers into her heart, and she even found herself feeling sorry for the singers, instead of feeling sorry for herself. Miyuki quite often caught herself wallowing in nostalgia for the days of hopeless yearning, and for the delicious balm of empathy that can only be felt in the emotional gutter, but she was never really serious about it. She knew the price she would have to pay to get those feelings back again.

She had long since let her membership of the fan club lapse, but as she read the book she found herself smiling almost wistfully at the times when she had done nothing but watch this actress's films for weeks on end, entire school holidays spent with the curtains drawn, sitting in front of video after video, lapping up the classics and even enjoying the ones where the hairstyles hadn't stood the test of time, and nor had a great deal else. This was as close as she could get to having fond memories of a past love, and it had been true love too, generous and pure. Years later her mother had told her it was this very obvious crush, which manifested itself in postcards and calendars, that had started her thinking that she probably didn't have the most heterosexual daughter in town (first position, she told her, had been awarded to a girl from the next street, who emerged red-faced from behind the bus shelter with such alarming frequency that by the age of twelve she was already widely regarded as being heterosexual to a fault). On the day a fifteen-year-old Miyuki had drunk a two-litre bottle of sweet cider and tearfully told everything, her mother had found herself quite ready for the revelation, and was more worried about having vomit on the carpet than she was about anything else. It turned out to be a valid worry, because Miyuki was indeed sick on the carpet. On the stairs too, where it was really hard to clean up.

Whenever she needed to, Miyuki could close her eyes and feel her mother holding her hair back from her face as she spewed the last of the cider into the toilet bowl. She could feel her rubbing her back, and hear her whispering over and over again that she loved her more than anything in the world.

In between chapters she looked up at a dog that was repeatedly chasing a stick into the freezing water, and at a pair of rubber-clad surfers as they paddled out on their boards before catching slow rides on low waves back towards the shore. In the middle of the afternoon she closed the book and started back the way she had come.

After a few minutes she heard the roar of an engine and the sharp blast of a horn. She looked around to see Mr Puw in his little flatbed truck. He was out on his rounds, delivering oil and gas to remote houses. He poked his head out of the window. 'Hop in, Thunderthighs,' he called. Miyuki didn't really feel like walking back, and was happy to take him up on his offer. She climbed into the cab, sat in the passenger seat and put on her seatbelt.

'Been busy?' she asked.

'Always busy.'

She didn't know what else to say, and was relieved when they rounded a bend in the narrow road to find their path blocked by Septic Barry's van. After a stand-off involving the blaring of horns and theatrically shaken fists, Septic Barry conceded defeat and pulled over on to the verge. As the vehicles passed he looked at Miyuki, twisted his face into an exaggeratedly horrified expression and shouted to Mr Puw from open window to open window, 'How long has this been going on?'

'That would be telling.'

'Just wait until your wife finds out,' said Septic Barry. 'She'll hit the roof. A man of your age, too – you ought to be ashamed of yourself.'

They drove on. 'He's a bloody prick, that one,' said Mr Puw, affectionately. Then he realised who he was talking to, and apologised for his language.

'That's OK,' she said, smiling. 'I've heard worse. And anyway, he is a bloody prick, isn't he?'

Mr Puw chuckled. 'Where do you want dropping off?'

'Anywhere's fine.'

He let her out on the Anchor side of the harbour, and she went up the main street and into the shop to stock up on a few provisions. The sun was just starting to head towards the horizon, and she had time for a quick walk along the coast path.

From the cliff top it was golden, just the same as before, and when she climbed down and approached it she shivered when she saw how well it had worked. It really was as if there was an enormous block of burnished gold on the beach. She could have stayed there forever, just gazing at it. She didn't, though. She didn't want anybody catching her there, looking so pleased with herself.

Her belly full of oven chips and barbecue sauce, Miyuki sat underneath the stuffed pike with her pint of Brains and her book. It was even quieter than usual. There were no walkers staying in the rooms upstairs, Septic Barry and the Children from Previous Relationships were in their smoky huddle on the other side of the pub, and short Mr Hughes and Mr Puw were standing in their usual places without saying very much at all. Septic Barry's girlfriend had the night off and had stayed in the caravan writing letters to friends, leaving Mr Edwards to pull the pints alone.

Nobody seemed to be mentioning the absence of tall Mr Hughes, but as the evening wore on and he didn't appear it started to bother her. She didn't want him to come in and reveal her secret activity by innocently telling short Mr Hughes and Mr Puw about their meeting that morning, but she worried less about that with every pint, and felt increasingly unsettled as she thought back to his gloominess. A sense of guilt crept along her spine and across her shoulders. She had been so concerned with getting off the beach that she had brushed him aside.

She wanted to know that he was feeling better, and to

listen to him for as long as he needed to talk, even if it was for hours and hours, which it more than likely would be. She kept losing her place in the book as she tried, unsuccessfully, to think of another evening when she had been in The Anchor and there had been no tall Mr Hughes. It seemed like a different place without his incessant baritone rumble reaching into every corner of the room.

The quiet was shattered when a group of young drinkers came in through the far door and stood at the bar talking loudly in Welsh. Although she could only make out the occasional word, she was pleased to hear the language being spoken, then straight away she felt like an idiot for feeling pleased. She wondered if she would feel similarly pleased to hear Spanish being spoken in a bar in Spain. She tried to transform her pleasure at hearing Welsh into dismay at hearing so much English, but she ended up just feeling annoyed with herself for being a virtual monoglot. She knew the basic civilities and insults and a few other bits and pieces, but the Welsh language hadn't been a big deal in her family or on her street, and she had never taken the time to learn it properly. On a good day her eavesdropping was passable, but she wasn't able to hold a conversation of any depth. Most of The Anchor's customers spoke in English, so it didn't provide her with too many opportunities to test herself. She had known tall Mr Hughes, short Mr Hughes and Mr Puw to drift into Welsh, and she understood just about enough to know that they were usually talking about the effects of the weather on their vegetable patches, which was how she had found out that it was a bad year for cauliflowers. It was never long before they moved back to English, though.

When she finished her sixth pint she went back to the bar.

'Holy mackerel,' said Mr Edwards, as he took her empty glass and pulled the pump.

Miyuki could tell that this particular *holy mackerel* was a comment on this being her seventh pint – it contained both a note of admiration for someone of her size having such a capacity for beer, and a gentle warning for her not to overdo it because she wouldn't be wanting a headache in the morning.

Mr Edwards was a man of few words, and most of these were *holy* and *mackerel.* He could load the phrase in so many ways. Depending on his tone and his manner it could be a greeting, a valediction, an expression of surprise, of pleasure or dismay, an admonition, a congratulation, a remonstration, or even a comfort in a difficult time. When Mr Puw's daughter, his only child, had died after years of illness, he had called Mr Edwards from a pay phone at the hospital.

'She's gone, Tristan,' said Mr Puw, abandoning their affectation of addressing one another as *Mr this* or *Mr that.*

'Holy mackerel, Bryn,' said Mr Edwards. 'Holy mackerel.'

There was more comfort in those words than in any of the cards of condolence that followed, or any of the expressions of sympathy from friends and relations. It said all the usual things you can expect to hear at such a time, things like *if there's anything I can do – anything at all*; and *she was a wonderful girl and we'll never forget her*, but it said so much more as well.

Mr Puw couldn't find the strength to reply. Without another word he put the phone back on its hook. As he made his way slowly through the corridors and back to his wife, it was as if he had a friend by his side.

Miyuki couldn't be sure whether Mr Edwards' *holy mackerel* had been an influence or not, but after her seventh pint she decided it was time to go home. Short Mr Hughes and Mr Puw were still there, and she told herself that if they weren't worried about tall Mr Hughes not turning up then she needn't be worried either. She stood up and walked over to the bar, where she put her empty glass down.

'I see some idiot's been painting rocks,' said short Mr Hughes.

Mr Puw grunted, and Miyuki felt as though a horse had kicked her in the solar plexus. She couldn't tell whether this comment was aimed at her, or if it was just a coincidence that short Mr Hughes had chosen this moment to mention that he had seen the results of her morning's work. Maybe he had already crossed paths with tall Mr Hughes and knew exactly who had done it, or perhaps it really was just his spontaneous and heartfelt response to her artwork.

Whichever way she looked at it, short Mr Hughes thought she was an idiot.

She left the bar without her usual nods of goodbye.

She chose a small log, and blew the fire back to life. Leaving the stove's soot-blackened glass door open, she stayed kneeling where she was, staring into the flames. She thought of Grindl and her bones ached, but this time she took no satisfaction from the melancholy.

Every year that she and Grindl had been together they had made a point of spending a month apart. This had been Miyuki's idea. She needed to know that they weren't going to become one of those couples who couldn't function without one another. Throughout her single days she had been exasperated by people who wore long faces when they were temporarily separated from their love interest. *At least you have somebody*, she would think to herself. When it became apparent that she and Grindl would be together for the long haul, she was determined for them never to be dependent on the immediate presence of one another for their happiness. Grindl had seen her point, and agreed to go along with her plan, so every year each of them would go away on her own for two weeks. These fortnights were governed by one simple rule: that there was to be no communication.

Without it ever being discussed, the reasons for them spending time apart quietly evolved. With the inevitable crises and jubilations, spats and reconciliations, she and Grindl had grown closer and closer, and more accustomed to one another's company. Miyuki still felt the need to know that they could cope with being apart, but these separations also began to take the form of a lesson in not taking one another for granted. She was sure that if they had one month apart each year they would appreciate each other all the more for the other eleven. So it was the aching bones and the quiet moments of desolation that told her she was still in love with Grindl. She already knew it perfectly well, but she still found it reassuring to have it driven home in this way.

Today, though, the feeling in her bones was different from usual, because along with the familiar ache from the absence of the one she loved was a sense of shame and embarrassment for having done something that she probably shouldn't have done.

She was jolted from her introspection by a sudden pair of sneezes, bringing her total to sixty. At this rate she would be dead long before the fortnight was up.

It was Grindl who had started her on this. Without ever talking about it, each of them had always made a point of saying *bless you* after the other one sneezed. Then one day, when an absent-minded Miyuki had failed to bless Grindl after witnessing a particularly elaborate sneezing fit that had run the gamut of sounds and expressions,

Grindl had looked horrified as she reached for a tissue.

Realising her oversight, Miyuki corrected herself. 'Oh,' she said. 'Bless you.'

'Just in time,' said Grindl. 'Another half a second and it would have been too late.'

'I didn't know there was a time limit,' said Miyuki.

'You have to say it straight away, otherwise it doesn't count.'

'And what if I hadn't said *bless you*? What would have happened then?'

'Nothing this time, because it was only seven. But what if you'd kept on forgetting and I'd reached a hundred? What then?'

'I don't know,' said Miyuki. 'What then?'

'I'd have dropped dead, that's what. Not that you'd have cared.'

Grindl couldn't believe that Miyuki hadn't heard of this. She explained that it was an established fact that anybody who was unfortunate enough to sneeze a hundred times with nobody saying *bless you* would find that their heart just stopped beating, and it would all be over. 'That's why people say it.'

From that time on, Miyuki had never failed to bless Grindl after a sneeze. She was always kept particularly busy during hay fever season, saving Grindl's life several times a day. Although she knew it to be a load of rubbish that even Grindl herself didn't really believe deep down, the superstition began to seep into her bones and she found herself counting her unblessed sneezes. She had always liked the idea of getting to one hundred, just to prove that it could be done without dying, but as her tally rose she found herself becoming just a little uneasy.

Maybe she really was just forty sneezes from the grave.

She opened her book and read the final chapter, with its inevitable account of the actress's early death. She tried to imagine what it must have been like for her to drown in the sea at night, to be dragged into the darkness as her lungs and belly filled with cold salt water. She had read that a drowning person's final moments are supposed to be imbued with a wonderful sense of tranquillity, but she found it hard to believe. She read the epilogue, with its inevitable platitudes about how her spirit lives on through her work and in the hearts of those who admire her. It wasn't the best way to end the day.

She closed the book, and gazed at the portrait on the front cover, at the brown eyes that looked straight back into hers as they had done so many times before. After a while she knew it was time for her to stop gazing, and she put the book on her *finished* pile.

She found a tissue, and blew her nose. She noticed that what came out was flecked with gold, and when she blew again still more gold appeared. She threw the paper into the stove and watched it flare up, blacken and disappear. After licking her right thumb and forefinger she plucked out her contact lenses, and dropped them on to the hot metal. She took no pleasure in watching them blacken as they danced a two-step before lying still, two small lumps of rabbit shit.

As she brushed her teeth she looked in the mirror and noticed flashes of gold in clusters of grey gunk at the corner of each eye. *Bollocks*, she thought, as she flicked them away with her flannel.

She had supposed she would be too wound-up to be able to sleep that night, but the last pint had been a good one. She got undressed, and before she had even got around to putting on her long johns and her long-sleeved T-shirt she got into bed. She closed her eyes, her jaw went slack, and she began a light, breathy whistle.

FRIDAY

Miyuki woke early with an unaccountably clear head, and couldn't get back to sleep. After staring into the darkness for a while, she got up. The electric heater's timer hadn't yet clicked on, and she reached a goose-pimpled arm through the cold air to the drawer by her bedside, then down to the floor. Her hand found her clothes and quickly pulled them under the covers, where they shocked her skin as she wriggled into them.

She braced herself, threw off the duvet and went into the kitchen, where she filled the kettle, took the biggest mug from the cupboard and made a black tea with two bags. While it was brewing she put on her boots and coat. Mug in hand, she left the cottage.

* * *

It was half past five, and the sky was black. The moon and stars were hidden behind thick clouds, but a solitary streetlamp cast a glow on the water and bathed the scene with a dull, jaundiced light. She sat on a low wall and let her eyes adjust. Her heart jumped when she realised she wasn't alone, that there was a figure on the quay, about thirty yards away. Although staying on one spot, it was very slowly melting in and out of the strangest shapes. She had always wanted to see a ghost, and thought her luck might finally be in until she remembered something Septic Barry had told her, and realised that the apparition was just the drummer of the Children from Previous Relationships doing his t'ai chi.

She had never spoken to him, and had no idea what his name was, but she was aware of his place in the scheme of things. Passing their huddle on the way to the toilet she had heard him talking about *pounding the skins,* and telling the others that he lived for his sticks. A few times she had seen him do the terrible thing that drummers do in pubs, tapping out a rhythm on the table with his fingers while sucking his cheeks, nodding and looking serious. Whenever he came up in conversation, Septic Barry referred to him only as *my drummer*, and Miyuki didn't feel the need to know a great deal more about him.

She watched as he slowly writhed around. Septic Barry had explained to her that even though his drummer had the self-discipline to get up and go down to the quay before breakfast every morning of the year, no matter what the weather, he had never taken a t'ai chi lesson in his life, and had learned all his movements from a long-returned library book. Miyuki found it hard to equate what she was seeing with a noble and ancient discipline. He seemed to be brushing the hair of invisible children, topping transparent eggs and repeatedly testing the tyre pressure of a bicycle that only he could see. She could just about make out the look on his face, and supposed that anything that could induce such an incredible display of grimacing must count as exercise on some level. He was mesmerising.

Since she had become aware of their existence she had wondered what it would be like to see Septic Barry and the Children from Previous Relationships playing live. She could tell just by looking at their hair that they wouldn't be much good, but even so she was curious and had occasionally asked Septic Barry if they had any dates lined up. He was always cagey about their concert schedule, until one night he crumbled beneath the weight of her polite inquiry, and confessed that they had never played in public.

'You can't rush these things,' he said, and the floodgates opened.

He told her the story of the band.

Around ten years earlier the village had been home to Impotent Rage, a group of teenage boys who specialised in Bon Jovi covers. When their bass player did unexpectedly well in his A-level retakes, gaining moderate passes in two subjects, he announced that he was leaving to go to college. While some of his band mates thought he had made a perfectly sensible decision, others considered his departure to be an unforgivable betrayal, and Impotent Rage collapsed. The ensuing inferno of accusations and recriminations resulted in several of the members vowing never to play music again.

In the wake of this implosion Septic Barry had sensed an opportunity. He rounded up his friends to form a group of their own, taking pole position himself and allocating each of the others an instrument, even though none of them had ever played a note of music outside their school xylophone lessons. With no more Impotent Rage, Septic Barry and the Children from Previous Relationships became the village band. Almost a decade had passed since their inception and they still hadn't made a public appearance, but it didn't matter because for that whole time they had been considered, by themselves and by everybody else, to be the village band, and that was more important than any such detail.

A pint into his story, Septic Barry also confessed that he and the Children from Previous Relationships had only ever had one rehearsal, in his caravan on a rainy afternoon three years after they had formed. It hadn't been a very productive twenty minutes, and since that day they had been content to discuss band business and their plans for the future in minute depth without ever doing a great deal about it. They named albums that had yet to have any songs, they discussed the possibility of getting sponsorship deals from instrument manufacturers, and they took turns in storming out after rows over musical differences before quickly re-joining, petrified that somebody new would be brought in to take their place. They worked out the best routes from venue to venue for a tour that would never be booked, they drafted a rider for their dressing rooms (soft towels, condoms, pork scratchings, four copies of the *Daily Star* and, most importantly, a barrel of Brains no matter where they were in the world) and they fostered intense rivalries with other bands, particularly Diamond Deluxe, a slick wedding outfit from two villages up the coast who they roundly dismissed for their lack of commitment to the true spirit of rock 'n' roll. Septic Barry and the Children from Previous Relationships did almost everything a normal band would do, apart from going to the trouble of playing live, recording or even practising together.

'We'll definitely do a concert one day though,' Septic Barry told her. 'Definitely one day.'

Over on the quay, Septic Barry's drummer looked as though he was milking an unusually long-legged cow, and was pulling the kind of face that might be expected of somebody whose testicles had become entangled in an electric fence. With a protracted series of contortions he pulled a cigarette from his pocket, put it in his mouth and lit it with a match. It hung from his lips as he continued twisting his body into inexplicable shapes. Watching the red glow in the darkness, Miyuki was reminded of the days when she would have a smoke before breakfast. Her memories were a little too fond, and she could feel a twitch in her cigarette fingers. She decided she had seen enough t'ai chi for one morning, and after sneezes sixty-two to sixty-four she picked up her empty mug and went back up to the cottage.

As she fumbled with the key she felt a drop of rain on the back of her hand, and hurried inside.

She spent the morning reading her book, the thickest one she had brought with her. In between chapters she listened to the rain, dozed, drank too much tea, ate handfuls of Flamin' Hot Monster Munch and occasionally tuned in to the weather forecast in the hope of hearing some encouraging news. She blew her nose a couple of times, and checked her eyes, and saw nothing unusual, but when her restlessness led her to stick a cotton bud into her ear she found that the anticipated yellow was mixed with tiny glints of gold.

Stir-crazy by lunchtime, she made herself as waterproof as possible and hurried through puddles over to The Boat Inn. A stream ran down the side of the road, and she walked through it, enjoying the feeling of the current against her Wellington boots. Her feet were OK, but the rest of her wasn't so well protected. Cursing herself for not having packed an umbrella, she was dripping wet by the time she got inside.

Once she had emerged from her coat and hat, the landlord turned white and said, 'Oh my God, it's you.'

He stared wide-eyed at her, his only customer, frozen apart from his Adam's apple, which bobbed up and down with alarming speed. After a while, he trembled back to life.

'I am so sorry about last year,' he said. It was something he had said to everybody who had come through his doors since the first of the month, whether he recognised them or not. It was better to be safe than sorry, he thought. He recognised Miyuki perfectly well though, and knew for certain that an apology was in order. 'Oh God, I'm sorry.'

In her hurry to get out of the rain, Miyuki hadn't looked at the blackboards by the front door. Usually they would be advertising lager promotions, or screenings of forthcoming rugby matches, but today they both said, simply, I AM VERY SORRY.

The Boat stood on the other side of the harbour from The Anchor. A cavernous place with fruit machines, a pair of pool tables, a jukebox and a big screen for the rugby, it wasn't really Miyuki's kind of pub, but she liked to go in once or twice during her stays, for the change of scenery as much as anything. Like at The Anchor, like everywhere, business was always slow at this time of year, and it wasn't unusual for her to be the only customer.

The Boat's landlord had a passion and a talent for passing the time of day, and it had been this that had led him towards his profession. Even after three decades in the business there were few things he enjoyed more than engaging in amiable conversation with the people who came in, be they regulars or passing trade, and he had noticed that amid all the talk about the weather, the tides, sport and local goings-on, a particular topic had arisen with surprising frequency.

Conversation at the bar of The Boat often turned to the subject of visits to other pubs, and the landlord would listen with genuine interest as his customers reeled off extensive lists of their favourite drinking places, and swapped recommendations and anecdotes. Regularly included on these lists were bars and restaurants where the staff were openly hostile towards the customers, to the point at which the atrocious service had become an attraction. There were examples of this kind of place all over the world, and apparently people would flock to them to be scowled at and insulted by grumpy waiters and dismissive bartenders. There seemed to be an implicit agreement between the customers and the staff, with the customers gladly suspending their expectation of smiling service and the staff giving them exactly what they wanted, their thunderous tempers and acid tongues every bit as professional as a polite manner would have been under normal circumstances.

Stories like these would animate the conversation like

nothing else, and the more The Boat's landlord heard, the more it seemed there was a considerable amount of mileage in this strategy.

I could do that, he said to himself.

When she had gone in the year before, Miyuki had thought she noticed a difference in him, but it was so fleeting and so unexpected that she hadn't quite been able to get her head around what happened. He had recognised her from her previous visits, and greeted her as warmly as ever, asking how she was, and how long she would be staying. When she ordered her pint of OSB he pulled it carefully, and placed it on the bar in front of her. Everything was carrying on exactly as she would have expected, until she asked for some dry roasted peanuts. Suddenly he remembered himself, and as he slammed the packet down on the counter he looked her straight in the eye and said, in a menacingly quiet and level voice, 'I hope they choke you, you fucking bitch.'

He immediately snapped back to his old self. He told her how much she owed, thanked her for her ten-pound note and carefully counted the change into her hand. Miyuki went to sit down, wondering if she had misheard him. She decided she must have done, that there was no way he would really have said something like that. Maybe she had had a fleeting funny turn, and imagined it all. Even so, she didn't feel like staying for another drink and she hadn't gone back there for the rest of her stay.

He had started his new strategy on the first of January, and when Miyuki had walked, mildly puzzled, from the pub, he had just been getting into his stride. His new persona quickly developed, and he became increasingly adept at spitting bile. An unsuspecting woman might, on asking for a glass of wine, be told that this wasn't a fancy wine bar, and that she wasn't going to be served a glass of wine, she was going to be served a pint of beer instead, and if she didn't like it she could, quite simply, take her piss flaps and fuck off. If she asked him why there were so many bottles of wine lined up on shelves behind the bar, he would explain that he put them there for the sole purpose of annoying people like her. Then, if a man was to feel peckish and decide to order a basket of chips to accompany his pint of beer, the landlord would inform him that he would be happy not only to provide him with his order but also to stuff it, wicker and all, right up his anus. Some people wouldn't even make it as far as the bar. On coming in through the front door they were greeted with 'Just turn around and get out, you pile of dog shit.' If they asked for an explanation, he would say that he felt nothing but bitter hatred towards them, that they were vermin and deserved to die – slowly and painfully.

Overweight people would be expelled on the grounds that 'Fat-suits are a fire hazard. What do you think you're doing coming in here with that thing on? Do you want us all to go up in flames?' Even customers he had known for years were told to stick their pickled eggs up their cock holes. Assuming he was going through some sort of personal crisis, they tried to remain loyal, but one by one they became so exhausted by his unremitting invective that they found themselves with no choice but to find new places to drink and to hope, from a safe distance, that sooner or later he would calm down and change back to his old self.

By the summer's end the pub was deserted, apart from the occasional passer-by who would walk in for a drink and a snack and be baffled by the landlord's livid face and his torrents of abuse. The anticipated upturn in business had failed to materialise, and he started to lose heart with his scheme. As Christmas approached, his insults became increasingly perfunctory. 'I don't think much of that shirt you've got on,' he would sneer at the stranger, or he might just curl his lip and say, 'Nice hair,' in a sarcastic voice. These people left after a single, hurried drink.

He was already on the verge of giving up when he read the new edition of a popular visitors' guide to their particular stretch of the coast path. Little more than a photocopied pamphlet, the guide came out every year and was sold in various places around the area. It was compiled by a semi-retired vicar from the Wye Valley, who loved the path and had a passion for introducing others to the various pubs, shops, boat trips, woollen mills, bed and breakfasts, and historical and geological curiosities to be found along the way. It never made him a penny, but that wasn't the point. He was always updating it, making discreet visits to a variety of establishments in the hope of finding new things to recommend to his readers.

Keen to avoid preferential treatment, he took care to ensure he didn't draw undue attention to himself as he went about his fact-gathering. He left his dog collar at home, and whenever he had cause to give his name he would say he was a *Mr*, rather than a *Rev*, and nobody thought for a moment that the man who was staying in Room 6, or sitting quietly in the corner of the tea room in an open-necked shirt, could be him.

His gentle nature prevented him from ever having bad things to say about the places he tried out. If a B.&B. had draughty windows and a violent cat, or if a surly girl in a café was to serve him a lukewarm cup of tea while talking to her boyfriend on the phone he wouldn't seek revenge in print, he would just tactfully neglect to mention the place and hope that standards would have improved by the time of his next visit. This is why his entry for The Boat, which tore this rule to shreds, had come to people's attention. A copy of the guide was put through the pub's letterbox between Christmas and New Year, and the relevant page was marked with a note that said, *Come along now, enough's enough. From a well-wisher.*

The landlord read the entry for his pub:

In previous editions I have been happy to recommend The Boat Inn as a friendly place to stop for a lunchtime drink or a simple meal. However, while researching this edition I went in to check up on it, only for the landlord to shower me with obscenities. For obvious reasons I am unable to transcribe them here. My crime was, apparently, having, and I quote, 'a stupid bald head'.

The vicar went on to withdraw his past recommendations of the pub, and to say that he hoped for a change of management in time for the next edition. If anything he had let The Boat off lightly. He had chosen not to relate the tale of how he had weathered the tirade and decided to stay in order to see if he could make any sense of what he was hearing. On handing him his pint the landlord, who had apparently recovered from his fit and returned to his customarily agreeable manner, suggested that the vicar have a smell of the beer.

The vicar took him up on this, and placed his nose just a fraction of an inch above the froth.

'Lovely, isn't it?' asked the landlord.

'Oh yes,' said the vicar, who enjoyed his real ale a great deal. 'It's very nice.'

'Smells like your wife's twat, doesn't it?' The landlord's eyes were fiery slits of hatred. 'Yeasty.'

The vicar, whose wife had died just a few months before, abandoned his drink and left the pub with shouts of 'I've had her, you know,' ringing in his ears. 'I've had her, and she was shit.'

During the New Year's Eve shift, in which not a single customer had walked through the door and he had stood alone behind the bar in a paper hat, The Boat Inn's landlord decided once and for all that the well-wisher had been right – enough was indeed enough. He resolved to change back to being a friendly publican from the stroke of midnight, and to apologise to everybody who came in from then on. A part of him hoped that somebody would turn up before midnight so he could bow out with a carefully chosen tirade, but nobody did. At one minute to twelve he walked into the gents' toilet, looked in the mirror and used it on himself.

'Just turn around and get out,' he spat at his reflection. 'If you seriously think that the first thing I want to see in the new year is your ugly face then you must be even more stupid than you look. You fucking tosser.'

He locked up, took off his paper hat and braced himself for a year of saying sorry.

Miyuki stepped out of the puddle that had formed around her, and accepted his apology. 'That's OK,' she said. 'Forget about it.' She wasn't really sure why she was forgiving him, but even so she was glad she had done.

'You don't know how much that means to me,' he said. The relief on his face gave her a warm glow, and she wished she had the guts to forgive people more often, to walk up to strangers in the street and say, *It's OK – it's just water under the bridge*, and watch the worry vanish from their faces as her absolution sank in. The more she thought about it, the more she wished people would do that to her too every once in a while.

She sat and drank her OSB, and ate a basket of deep-fried potato wedges which she smothered in sachet after sachet of salad cream. She read her soggy book, trying not to let the pages disintegrate as she turned them, then went back to the bar to order another pint and replenish her depleted supply of salad cream sachets. When she decided she had drunk enough for a lunchtime she quietly put her empties on the bar, and wriggled back into her damp coat as the landlord apologised to a pair of drenched walkers who had just come in.

'I'm so sorry,' he said.

'What for?'

'Last year.'

'What about last year?' This was the first time either of the walkers had ever visited the village, so they had even less idea than Miyuki of what he was on about.

'I'm just sorry, that's all. Can't you forgive me?'

'OK then,' said one of the walkers. 'We forgive you.' For all they knew they were forgiving him for snatching an old lady's purse on pension day, or for rubbing up against a schoolgirl in a swimming pool, but it had been a horrendous morning and they were both desperate for a drink.

As Miyuki stepped back out into the rain, she heard them order two pints of OSB and two packets of plain crisps.

She woke to find that the rain had more or less stopped, and the clouds had turned from dark grey to light grey. There was still some daylight left, and she couldn't resist going to the cliff top to see her golden rock.

Without the sunshine bouncing off it, it didn't have quite the same dazzling brilliance to it as the evening before. She made her way down the path for a closer look. The ground was slippery, and it took her longer than usual to get to beach level. As she got closer, she noticed there was a smear of rain-sodden seagull shit across it. Even so, it was still quite something. The cloud cover and the approaching nightfall made the world appear almost monochrome, but the gold leapt out from the grey as though it was the only real colour left in the world.

She told herself that short Mr Hughes didn't know what he was talking about. It looked magnificent, and nothing at all like the work of an idiot.

Scrabbling and sliding her way back up the path, she remembered at last to wonder if she would run into tall Mr Hughes that day. She was sure she would.

Even though it was Friday, The Anchor was having another quiet evening. There were still no walkers staying in the rooms upstairs, and Septic Barry and the Children from Previous Relationships were confidentially discussing band business at their usual table, all of them sucking on cigarettes apart from Septic Barry, who had given up because he was saving his voice for the big gigs. Short Mr Hughes and Mr Puw stood at the bar, only occasionally saying anything to one another, and Septic Barry's girlfriend was passing the time by dusting under bottles. The rain had come back, and was pelting against the windows.

It wasn't until Miyuki's third pint that she overheard short Mr Hughes and Mr Puw talking about tall Mr Hughes.

'He's not in again,' said short Mr Hughes.

Mr Puw nodded. 'You seen him around?'

Short Mr Hughes shook his head. 'He'll be up to something, no doubt. Looking for God's nostril again, I expect.'

The last time tall Mr Hughes hadn't turned up to the pub he had been in the following evening with an elaborate explanation for his absence. He told them he had been at his kitchen table doing a 2,000-piece jigsaw puzzle of the ceiling of the Sistine Chapel, and had become so engrossed in looking for the piece with God's nostril on it that he had completely lost track of time. He repeated this anecdote every once in a while. *I looked down at God's nostril*, he would say, *and then I looked up at the clock and it was already the next day*.

'He'll be in tomorrow,' said Mr Puw. 'He's not going to miss three nights in a row.'

Short Mr Hughes turned down the corners of his mouth, nodded and went back to his pint.

Miyuki didn't understand the reference to God's nostril, but it was clear that they hadn't seen or heard anything of tall Mr Hughes. She told herself she didn't have to worry, that he was a grown man with his own life and he wouldn't appreciate people making a fuss about him. And besides, it wasn't as if he was missing, it was just nobody had seen him for a while. That was all.

Without tall Mr Hughes drinking alongside them, short Mr Hughes and Mr Puw were free to say anything they felt like saying about him. 'Do you remember that time when . . .?' asked short Mr Hughes, finishing his sentence by casting a theatrical look past his moustache and down to his fly.

'Oh, yes. I remember, I remember,' said Mr Puw, in a tone that told short Mr Hughes that this wasn't the time or the place for such a reminiscence, that the Japanese girl and Septic Barry's girlfriend were in the vicinity, and the story wasn't at all suitable for the ears of young ladies.

Miyuki overheard this short exchange too, and again she couldn't stop herself from wondering what they were talking about. All the years she had been going to The Anchor she had been happy to fade into the background, and be more or less oblivious to everything that was going on around her, but the absence of tall Mr Hughes made her wish she had just a bit more of an idea of what people were talking about.

She took a long drink of her beer and went back to her book, oblivious to the discomfort of the two men standing at the bar.

Short Mr Hughes had unleashed the image, and he and Mr Puw stood in silence, unable to stop themselves from recalling in appalling detail the evening he had been alluding to, when tall Mr Hughes had made a visit to the lavatory and neglected to put his penis away before returning to the bar. He had picked up his pint and launched straight back into a monologue about bridges that had started two nights earlier. It had been *bridges* this, and *bridges* that. They found it impossible to concentrate on a word he was saying, which wasn't unusual, but this time it wasn't because their minds had been numbed by the monotony of it all, it was because there was a penis dangling before them.

Short Mr Hughes, Mr Puw and Mr Edwards didn't hear a thing as tall Mr Hughes continued his summary of the life and work of Thomas Telford. They tried not to look, but it just wasn't possible. It was as if they were hypnotised. None of them had ever thought to wonder how big it might be, not even for a moment, but independently of one another they concluded that if they had done they would have never have supposed it to be anything like the extraordinary length or girth of the veined beast hanging right before their eyes.

It was Mr Edwards who put a stop to the situation,

with a nod of his head in the direction of the oversight, and a well-timed *holy mackerel.*

Tall Mr Hughes looked down, and saw it.

'I knew that would happen one day,' he said, and he put it away. He adjusted himself, then zipped his fly as he carried straight on with explaining to deaf ears all about the difficulties of spanning a raging torrent. 'Imagine,' he said, as if he hadn't been exposing himself just moments before, 'laying foundations down there, with all that water rushing by. And that was in the olden days, too. They didn't have all the fancy tools you get today. You've got to hand it to them.'

No nickname had come from this faux pas, and short Mr Hughes's suggestion that he and Mr Puw reminisce about it was as close as anyone had come to mentioning it since. As they stood in silence at the bar, the pictures in their minds' eyes alarming in their clarity, short Mr Hughes felt embarrassed for having brought it up. He was relieved when the thermostat on the refrigerator clicked and sent the motor rattling unhealthily into life, because it provided him with an apparently natural opportunity to move the conversation in a new direction.

'That fridge needs looking at,' he said.

Mr Puw nodded, and relit his pipe.

Miyuki had been reading the same page for ages without taking in a word. She put the book down and stared at the watercolour dray horse on the brewery calendar, wishing she knew how she ought to feel. Again, she told herself she should be taking her cue from the men standing at the bar – they had known him for years, and if they weren't too worried about him then she shouldn't be either. Even so, tall Mr Hughes kept re-appearing in her thoughts, and whenever he did she was surprised at how vivid he seemed. It was as if his absence had somehow made him three-dimensional for the first time.

She read a few more chapters, drank more beer and gnawed her fingernails. At around ten o'clock Septic Barry came over to sit beside her. They shared a packet of scampi flavour fries, and she listened as he talked about some of the news stories he had read in the latest edition of the *Daily Star*, and told her some anecdotes relating to the disposal of human waste.

Then, just as he neared the bottom of his drink and his girlfriend was in the cellar changing the barrel of Brains, he asked Miyuki, quietly, 'So, do you think she likes me?'

'Yes, she likes you. You don't have to worry about that.' All evening Miyuki had been noticing the way Septic Barry's girlfriend's eyes shone when she talked to him, and the way she smiled to herself whenever she looked over at him as he huddled with the Children from Previous Relationships.

'The thing is, I know she likes me, but what I mean is, does she *like* me? It's not always easy to tell, and you know . . .' He couldn't look her in the eye. 'Well, I've got to know for sure, because I don't want to make an idiot of myself.' As he rubbed his palms up and down his jeans, Miyuki could see that the strutting ladies' man had been replaced by an awkward, shuffling teenager.

She smiled. 'Yes, she *really* likes you. It doesn't make any sense, but she does. Trust me, I can tell.'

'And do you honestly think I've got potential?'

'Yes, I do.'

'You're not just saying that?'

'No.' She was surprised to realise that she was telling the truth. 'No, I'm not. I really do think you've got potential.'

'That's good to know. Thanks for that.' He went back to sit with the Children from Previous Relationships.

Miyuki had a fair idea of what Septic Barry had in mind, and why he was so fearful of saying, or rather asking, the wrong thing. She smiled at the thought of it, and had a feeling it would work out well for both of them.

She finished her book, put it on the table and started daydreaming about Grindl, who by this time on a Friday night would be in the pub with her fellow volunteers from the local youth club. Every once in a while Grindl talked Miyuki into joining them, and she would sit there drinking her beer guiltily, as if she hadn't earned it. She had never stepped foot inside the youth club, choosing instead to spend her Friday evenings slumped in front of the television, so she mainly just sat and listened, a bit embarrassed for being there, and feeling like something of an interloper as Grindl and her friends exchanged the evening's tales of teenage indiscretions, and tried to get as many well-deserved drinks inside them as they could before last orders.

When Grindl had started volunteering, Miyuki had admired her for her social conscience and her commitment to the community. It took a while for her to realise that she didn't go there out of a sense of duty, or because of any higher motives, she went there because she enjoyed it. She really liked spending her free time with the pubescent boys in baseball caps who had joyously shouted *Lesbians!* at them in the street when they moved into the town. Once she had established herself at the youth club, the boys would still joyously shout *Lesbians!* at them in the street, but now they knew it would be the beginning of a long exchange of banter from which Grindl would emerge victorious. Watching this happen, and staying out of it, Miyuki saw that she really was fond of them, and that they were fond of her too.

She knew she didn't need to have Grindl's excellence affirmed by a gaggle of smoking, spitting youths, but as she watched her talking circles around them it was as if the rest of the world had receded, leaving just Grindl and her hawking cherubim, bathed in a golden light.

She shuddered out of her reverie, and lifted a hand as short Mr Hughes and Mr Puw nodded their goodbyes.

They seemed to be in perfectly amiable moods as they wrapped themselves up in their coats and hats, so she supposed they didn't know it was her who had sprayed the rock. They were followed by the Children from Previous Relationships, who slunk out of the PUBLIC BAR door, and by ten past eleven she was the only customer left apart from Septic Barry, who was waiting for his girl-friend to finish cleaning the drip trays.

Miyuki drank up, took her glass back to the bar and said goodnight to the two of them. Septic Barry flashed her a glance that told her to keep their earlier conversa-tion to herself, and she reassured him of her complicity with a smile.

It was a mild night, and she didn't feel the need to put a log on the fire. She turned on Radio 3 and listened to a recording of a woman from Bolivia making a kind of squelching sound. She plucked out her contact lenses and dropped them on to the top of the stove. It still had just about enough heat left in it to send them into a lacklustre fandango, and at points the sound of their slow death seemed to harmonise with the Bolivian squelching, but it wasn't long before they lay still and quiet, looking like nothing but a pair of partially shrivelled contact lenses.

She turned off the radio, brushed her teeth, checked her eyes and ears, and blew her nose. She was relieved to see that her body seemed to have stopped producing gold. A minute later she was in her long johns and her long-sleeved T-shirt, and she pulled the covers over her head.

From deep inside the duvet she let out a sigh, followed by a series of breathy grunts, and then there was quiet.

SATURDAY

Forecasts of a clear weekend had brought people out to the coast path. Every few minutes Miyuki found herself nodding a greeting to a passing dog walker, stepping gently around a prostrate amateur wildlife photographer, or exchanging positive remarks about the weather with a smiling couple in co-ordinated fleeces. Her solitude shattered, she left the main trail behind and took a narrow path that ran inland and up to a high crag.

At the summit she caught her breath, and felt goose bumps as she looked out over the shoreline and the moorland, and across miles of fields to the hills that she had sometimes thought about visiting. Without a trace of haze the greens, blues and reddish-browns were as vivid as they were ever going to be. The tall, skinny chimneys of the distant oil refineries and the occasional patch of caravan site tried to ruin the view, but she wouldn't let them. The landscape was peppered with civilisation, with standing stones, farm buildings, Neolithic burial chambers and dry-stone walls. The more recent additions were just another phase the landscape was going through. On a day like this she could even convince herself that they had a kind of beauty.

A kestrel hovered above her. She hoped to see it swoop, but it didn't. It drifted away and she followed its lead, walking downhill and along the paths through the fields with a familiar destination in mind.

The walk took longer than she had anticipated, and by the time she reached the pub the cold had bitten into her and she was thirsty for a winter ale. They still had a barrel of Cwrw Santa, so she ordered a pint and a packet of dry roasted peanuts and sat at the table by the fire. Three men and a woman were talking at the bar, and she tried to listen in. She could just about get the gist of the conversation, but that was all. Unlike The Anchor, in this pub it was the English language that only cropped up occasionally.

The further she strayed from the tourist trail the more Welsh she heard, and this wasn't the kind of village that attracted visitors, not even at the height of summer. It wasn't really on the way to anywhere, it was too far from the sea to lure in passing walkers, and apart from a few old farm buildings the houses were modern. For miles around, whenever an old stone cottage came up for sale it would be snapped up at an inflated price by somebody who would rent it out to holidaymakers. The cottage Miyuki stayed in was one of these, and when she saw the nondescript buildings that had been thrown up for the displaced locals she sometimes felt a shiver of colonial guilt. The planning department must have decided that the only way for people with normal jobs to be able to afford to keep on living in the area was to build a large proportion of new homes in a style so boring to look at that no tourist would ever want to stay in them.

The pub was plain too, a brick building that could have gone up at any time over the last hundred years. She liked it well enough though, and had been in quite a few times before. She stared into the flames, and as she tuned out the voices she found herself, as she often did, dwelling on her failure to properly learn the Welsh language. Grindl was from a bilingual family in a Welsh-speaking area, and she was always offering to teach her. They had tried a couple of times, but Miyuki had felt too self-conscious for it to be a success. Whenever Grindl was obliged to correct her it was an ordeal for both of them, and Miyuki's will to continue drained away. She always called a stop to the lessons before any real progress was made.

She resolved to try harder, and when she finished her beer she decided to order her next one in Welsh. This was basic, and she knew how to do it, but when she got to the bar her confidence evaporated and the words came out in English. Deflated, she sat back down with her fresh pint.

She started on her book, and was vaguely aware of a couple more customers arriving. They spoke in Welsh too, but their conversation seemed to be technical and she tuned out. It wasn't until she finished the chapter that she looked up and saw the inimitable hair of Septic Barry.

She couldn't remember ever having heard him speak a word of Welsh, but there he was, in full flow. When his drinking partner went to the gents, Septic Barry noticed Miyuki and came over to say hello.

'Out for a walk?'

She nodded. 'How about you?'

'Oh, you know. Doing a bit of this and that.' He tapped his nose.

'OK, I won't ask. Are you going to be in the pub tonight?'

'I suppose I might make it in for a half at some point,' he said. Miyuki took this as a firm *yes*. The other man returned to his drink. 'Back to business,' said Septic Barry, tapping his nose again.

Miyuki tapped hers back at him.

'It never ends,' he said. 'I'll see you in The Anchor then, Goldfinger.'

She watched him go. *Goldfinger*. That was a new one. She started a new chapter, and the next time she looked up they had both gone. She finished her drink, went to the toilet and stepped back into the cold day.

* * *

Embarrassed by the territorial feelings that had driven her up the hill, she snaked her way through the fields to the coast path for the walk back. Once she got there she tried to feel warmly towards the people she passed, and found it wasn't as hard as all that. They all seemed perfectly nice.

The light was perfect, and she almost wished she had packed a camera. For the first few years she had returned with a couple of rolls of film, but had given up when she found she was giving Grindl exactly the same slideshow over and over again. The cliffs didn't change from year to year, the village looked the same and there were always a few wild ponies who looked uncannily similar to Bryan Ferry.

In some places the path was narrow, and ran close to the cliff's edge. It had become something of a ritual that before she went away, Grindl would make her promise not to fall off. Walking holidays weren't Grindl's idea of fun – she preferred getting lost in strange cities, and when it was her turn to spend two weeks away she made a point of going somewhere different every year. Sleeping in dormitories and on friends' floors, and eating in markets and from street stalls, these stays in San Francisco, Athens or Marrakech cost about the same as Miyuki's drink-sodden fortnights less than a hundred miles from home. Miyuki could have travelled the world too if she had wanted to, but she couldn't think of anywhere she would rather be than the coast path on a cold January day, with peanuts in her belly and beer in her veins.

She stopped and looked down. The drop was sheer, and she couldn't see the ground directly at the bottom. She was glad she had made her promise to Grindl, and wondered how long it would take for her to be discovered if she was to fall. If it was the kind of cove that filled with water at high tide her body would drift on the sea, maybe coming back in at some point, or being caught in a trawler's net. Or maybe it would be sucked into the deep, and never be recovered. If she landed above the tide line she could lie there for months, hidden from the walkers on the path above. Her remains, picked clean by the gulls, would eventually be found by an abseiler or a canoeist. She quite liked the idea of being a skeleton one day, but she hoped that wouldn't be happening for some time yet. As if she needed a further memento mori, a raven left its perch on the cliff face and soared up in front of her, not much more than an arm's length away. She leapt backwards, and her heart thumped as she watched it fly away.

When she got back to the village the dusk had set in, her legs were aching and she was ravenous. She couldn't face the thought of heating up her own food, so she went to the café and ordered fried egg, chips and beans. While it was cooking she opened a copy of the local paper that somebody had left behind, and scanned the news. In the towns a drunk had been arrested for breaking the nose of an unfortunate passer-by, a lovelorn soul had expressed his heartache by setting fire to his ex-girlfriend's car, and a dawn raid had caught a pair of small-time drug dealers. Elsewhere some roadworks had gone on longer than expected, and a soldier who looked about twelve was leaving for the Gulf. There were reports of charity stunts, controversial planning decisions and rowdy public debates about wind turbines, and towards the back, in between the readers' letters and the sport, she noticed a column carrying a round-up of items from the village. When she saw the main headline she covered her eyes, and it took a while before she could read on.

Goldfinger Moves South?

Police are investigating after a rock was defaced with gold spray paint in the early hours of Thursday morning. They believe this may have been the work of the infamous 'Goldfinger', who in recent years has painted several rocks on beaches in Cardigan Bay. However, they are not ruling out the possibility of it being the work of a copycat vandal. A police spokesman said, 'Goldfinger is a menace, and has to be stopped.' Anybody with information is asked to come forward.

The next item was headed *Hospice Raffle*, but her eyes had lost focus and she didn't read on. She knew the paper came out on Fridays, and it would have gone to press on the Thursday so her efforts must have been noticed and reported to the police and the press more or less straight away.

When her food arrived she ate too fast, hardly tasting anything, and as soon as her plate was empty she got up to leave. The man behind the counter smiled and thanked her, and she thanked him in return, wondering whether he would have been so friendly if he had known he had been feeding a defacer of rocks, a menace, a copycat vandal.

Going into the shop for a loaf of thick-sliced white bread, she sniffed and felt a dull throb behind her eyes. Worried that she had a full-blown cold coming on, she decided to buy an apple in the hope that some fruit would stave it off.

She pulled off her socks and sat in the armchair, watching the flames and warming her toes as she drank a cup of tea. She couldn't be bothered to shower, and after a few pages of her book she started thinking about getting ready to go to the pub.

Her preparation for her nights in The Anchor rarely extended beyond checking she had her beer money, her book and her door key, and glancing in the bathroom mirror to make sure nothing was dangling from her face. This time though, rather than just glancing she stared at her reflection, at the face of a wanted woman. Septic Barry knew, and tall Mr Hughes knew, and there was a strong chance that everybody knew, so she would just have to accept the consequences. If it came to it she would admit everything, and as a remorseful first offender she would probably escape with a caution, or at worst a fine or a bit of community service. Whatever happened, she knew she should meekly accept her fate. The most important thing would be to keep it all as quiet as possible, in the hope that Grindl's mother would never find out.

For a long time Grindl's mother couldn't find a good word to say about Miyuki. On one visit in their early days she had taken her daughter aside for one of her *quiet words*, and told her that she thought Miyuki was *not as bad as all that, in spite of everything*. At that point this was the nicest thing she had ever said about her. Despite this jolt in the right direction, progress had remained slow. On a more recent visit she had taken Grindl aside and said, 'Miyuki's a perfectly nice girl, but I long for the day you stop all this *lesbian* nonsense. You can still be friends, you know. Will you think about it? For me?'

Grindl nodded, and told her she would think about it. On the drive home she told Miyuki about this conversation, and explained that she hadn't been lying. She knew she would be thinking about it a lot, just not in the way her mother had meant. She could never bring herself to tell her mother how much her quiet words hurt.

Even if Grindl's mother never found out about it, if her rock painting was to become public knowledge she knew she would have to leave the village in disgrace and would never be able to return. She wouldn't be able to sit and drink beer and read her book at the round table in the corner of The Anchor, underneath the stuffed pike, and that would be the real punishment. Her head throbbed at the thought of it, and after banging a clenched fist against her skull she threw a log in the stove, closed the door and went to the pub, hoping it wouldn't be for the last time.

The pub was lively. A group of walkers were playing darts in the public bar, and the rest of the customers were sitting or standing in groups, discussing the sports results or dissecting the day's surfing. She was relieved to see that she hadn't attracted any particular attention; nobody looked as if they were gearing up to wrestle her to the ground and restrain her until the police arrived. Short Mr Hughes and Mr Puw were standing in their usual places in the lounge bar, apparently untroubled by the absence of tall Mr Hughes. Maybe they knew he was fine.

She avoided eye contact with the stuffed pike as she took her pint over to the table in the corner. The voices merged into a drone, and she went back to her book. By half past nine the place had quietened down, and she looked up to see Septic Barry arriving at her table with a pint of Brains in each hand.

'Lesbians,' he said, gleefully bypassing any civilities as he sat down beside her.

'What do you mean *lesbians*?' asked Miyuki.

'In the *Star*. You know the photo story on the problem page? It's been running all week and I had a feeling it was building up to that. It's called *Is This Wrong?* and it's about these girls who take their friendship to the next level. In the shower, too. I love stuff like that. It's the last instalment tomorrow. Can't bloody wait. Have you ever met any lesbians?'

'Oh, one or two,' she said. 'One or two in my time.'

'I don't know any. Shame, really. There should be a lot more of that kind of thing going on, if you ask me.'

Miyuki found herself entertained and fatigued in equal measure by men's endless fascination with lesbianism. It seemed as though there wasn't a straight man on earth who wouldn't loosen his tie at the thought of a pair of girls rolling around together. She was tempted to tell Septic Barry that he was lucky enough to be sitting next to a real-life lesbian, but she decided not to. She had her rule, and besides she wouldn't want to sully his *Daily Star* fantasy by dragging it into the real world. It would be like telling a five-year-old that there was no such person as Father Christmas.

He stared into space for a while, before casually asking, 'Seen Hughesy lately? The tall one?'

'No,' said Miyuki. 'Not for a while.'

'He'll turn up.' He looked at his pint. 'Can you let me know if you see him, though? I need to ask him something.'

'I will do. And can you tell me if you see him too? Just so I know I don't have to look out for him any more.'

He nodded. They sat in silence for a while. She hadn't realised that Septic Barry and tall Mr Hughes had ever paid one another the slightest attention, but for some reason he seemed to be really anxious to contact him.

'Have you asked those two?' She nodded towards short Mr Hughes and Mr Puw.

'I asked them on the way over. They've not seen him for a bit, either.'

Miyuki's heart sank.

'He'll turn up,' said Septic Barry, and he started talking about some backed-up sewage he had dealt with that morning, how it had oozed into a bidet. It was a long anecdote. Miyuki had difficulty concentrating on it.

When it was finally over she said, 'I'm not Goldfinger, you know.'

Septic Barry gave her a look.

'I'm really not.' He gave her another look, and she knew she owed him as straight an explanation as she could give. 'I did spray the rock, but I'm not Goldfinger.'

'Oh, right. So you're a copycat vandal, then?'

'Well, no. I didn't know there was anyone else out there spraying rocks, so I wasn't copying. And besides, I didn't think of it as vandalism. I suppose I thought of it as art.'

'Modern art, is it?'

'Kind of. Well, no, not really. I just wanted it to be nice to look at. Honestly, Septic Barry, I would tell you if I was the real Goldfinger.'

He accepted this with a nod.

'Who else knows it was me?' she asked.

He smiled. 'I'm not a grass.'

She was relieved to hear that. 'But how did you find out about it? Did you see me do it?'

'No, you had gold in the corner of your eye that night. I could see it shining from the other side of the bar. I didn't think much about it at the time, but when I heard about the rock I put two and two together. That always happens to me when I spray my van, see. It gets up my nose and in my eyes and I end up looking like a bloody drag queen.'

Miyuki marvelled at the thought of Septic Barry in drag. It was hard to see beyond his haircut, but whenever she did, she noticed he had nice features. He was actually quite handsome, and might not look too bad in lip gloss and feathers. 'But what about all the others? Wouldn't they have noticed it too?'

'What, short Hughes and Puw? I don't think so. And even if they did, they probably just thought it was make-up. Same with that lot,' he said, pointing a thumb in the direction of the Children from Previous Relationships. 'They don't know women like I do.'

She wanted to give him a friendly hug, but she knew she couldn't. 'Thanks, Septic Barry,' she said.

'What for?'

'For keeping this to yourself.'

'Any time. After all, we all have our secrets, don't we?' He tapped his nose again. 'It's a shame really, though. I liked the idea of you being Goldfinger.'

'Sorry to be such a disappointment.'

He shrugged. 'Can't win them all.' As he stood up to leave, he said, 'I went to see it yesterday as a matter of fact, after I read about it in the paper. I quite liked it. It looked pretty good.'

Miyuki smiled, and watched as Septic Barry headed to the other side of the pub, where he walked past the Children from Previous Relationships and sat with Blind Billy and Blind Billy's wife. Miyuki had seen them coming in. She remembered Blind Billy from some of her previous visits. One evening Septic Barry had told her all about him.

It was clear that the child's eyes hadn't developed as they ought to have done, and when the doctors confirmed that there was nothing they could do to put them right, his parents made a decision. They had been planning on Richard for a boy, but instead they called him William. 'It'll be nice for him to have a nickname,' the proud father explained to all and sundry, as his wife nodded her agreement. 'I've always enjoyed having one. Jonesy, they call me – it really lifts my spirits.' When they sent the announcement cards, they read:

IT'S A BOY!
WILLIAM 'BLIND BILLY' JONES.
7LB 2OZ.

Blind Billy had grown up next door to Septic Barry. He was away at a special school a lot of the time, but he came back for weekends and holidays and the boys spent a lot of time together. To Septic Barry, Blind Billy's blindness was something that had always been there, but even so he would dwell on it from time to time, lying on his bed with his eyes shut tight, wondering what it would be like if they didn't work at all. One time his curiosity had led him to make a blindfold out of his school tie, and after a few hours of bumping into things as he charged around the house and garden, a bruised but exhilarated Septic Barry came to the conclusion that it really wasn't as big a deal as all that. He and Blind Billy always found a way of having fun together, and that was all that mattered until the day he stepped outside to find himself in a whole new world.

Before him was a magical panorama of cascading hair, of swinging hips and upturned noses, of shy smiles and confident strides, of slim waists, thick waists and breasts of all shapes and sizes. Sights that had been insignificant to the point of invisibility had, without warning, become explosions of wonder. Within seconds his reason for being had changed beyond recognition, and when Blind Billy got home that weekend Septic Barry had no idea what to say to him. Until that point they had shared everything, but there was no way they could ever share this, and for years it remained something that they both knew lay between them, unacknowledged. He had walked into a wonderland into which Blind Billy could never gain entry, and for the first time in his life he felt truly sorry for his friend.

Blind Billy ended up moving away for work, but he came back to the village every few weeks and always ended up drinking in The Anchor.

Convinced he would turn out to be some kind of virtuoso, Septic Barry had fruitlessly expended a lot of energy trying to recruit him into the Children from Previous Relationships. Eventually, though, he had to concede that Blind Billy wasn't going to join. He had turned his back on stardom, and although he couldn't begin to understand it, Septic Barry knew he had to respect his decision.

On one of his visits to the village Blind Billy had arrived at The Anchor as usual, but instead of sitting where he usually sat, in among the Children from Previous Relationships, he went straight to the other side of the pub. And beside him, drinking bottled lager and touching his leg, was a girl.

Septic Barry and the Children from Previous Relationships were enthralled, and wondered how on earth Blind Billy had been able to select her, of all girls, to be his girlfriend. Later in the evening he brought her over to meet them. They listened carefully to her voice, which seemed quite normal. Then each of them tried to smell her. It wasn't easy to do this without it appearing suspicious, but as they took turns in bringing the drinks over from the bar they leaned in front of her as they put the glasses down, and on drawing back they got their noses as close to her as possible, inhaling deeply. She carried a light scent from a shampoo or a lotion, but she really didn't seem to smell a great deal different from any other girl. They kept their hands to themselves, but could tell just by looking at her that her skin was soft and smooth, and they decided that Blind Billy probably had quite a nice time rubbing himself against her. None of them tried to lick her to find out how she tasted, although they wouldn't have minded because somehow he had managed to choose an incredibly good-looking girl. The longer they looked at the shape of her body, and imagined the texture of her skin, the less sorry they felt for Blind Billy, and while they were pleased for him to have found a girlfriend they also felt slightly aggrieved. It almost seemed as if he was being a dog in the manger: he could just as well have found himself a plain girl who smelled nice and had soft, smooth skin, and who didn't wear such a short skirt, and left this vision for a man with working eyes.

Blind Billy must have known he was on to a good thing, because a few months later he asked her to marry him. She accepted, and Septic Barry was appointed best man. Under normal circumstances he would have taken the groom-to-be to a strip club for his stag do, but instead they went on a surprise trip to a cheese factory, because he knew how much Blind Billy liked the smell of cheese. And afterwards they went to a pub, where they drank until they threw up.

Blind Billy's girl was the best-looking bride Septic Barry and the Children from Previous Relationships had ever seen. Her dress was tight and low-cut, and at the reception Septic Barry, who by this point had drunk a whole bottle of Penderyn whisky, delivered his entire speech to her cleavage. At one point, spellbound by the magnificence before him, he lapsed into a long silence and had to be prodded back to life by the father of the bride.

And while they were mostly glad for Blind Billy on his big day, Septic Barry and the Children from Previous Relationships still couldn't help feeling just a bit put out.

Two children later she was just as attractive as ever, but now, as Septic Barry sat with them in The Anchor, he looked over at his own girl as she pulled pints, and for the first time he was able to feel unreservedly happy for his friend.

'You're a lucky bastard, Blind Billy,' he said, when Blind Billy's wife was in the toilet. 'A very lucky bastard . . . and so am I.' They clinked glasses, and drank to it.

The pub had all but emptied, and Miyuki was on the last few pages of her book. She hadn't really been able to get into it, and wished she had chosen non-fiction that morning instead. Sometimes she just wasn't in the mood to read about the comings and goings of imaginary people, but she carried on. Mr Edwards was upstairs counting the takings, and Septic Barry's girlfriend, finished for the night, sat at the far end of the bar with a glass of white wine and a cigarette, talking to Septic Barry. The rumble of conversation had been replaced by the unhealthy mechanical whirr of the refrigerator behind the bar, and their voices were muffled. Then there was a click and a thud. The noise stopped, and Miyuki could make out every word. She tried to lose herself in her book, but she wasn't able to. Glancing up, she saw Septic Barry take his girlfriend's hand.

'There's something I want to say.'

Oh no, thought Miyuki, *I shouldn't be here for this,* but she weighed up her options and found there was nothing she could do. If she drank up and left they would notice, and wave goodbye, and the magical moment would be shattered. Maybe Septic Barry would lose his nerve, and it would be her fault. She brought her book close to her face, and pretended she couldn't hear a thing.

'I've been doing a lot of thinking lately,' he continued. His speech had obviously been rehearsed. 'I've been thinking about the future and what it holds.'

Miyuki tried not to listen, but it was no use, and she couldn't stop herself from casting a glance at them. Septic Barry's girlfriend could barely contain her delight. Her eyes were burning with anticipation.

'And I've decided . . .' he said, clasping her other hand and looking straight into her eyes, ' . . . I've decided that I'm going to expand. As of this summer I'm going to be providing temporary toilet facilities for small-to-medium-sized public and private events.'

Miyuki peered over the top of her book, expecting

Septic Barry's girlfriend to be looking like thunder. But she wasn't. She was still smiling, and without missing a beat she raised her fingers to his cheek, and said, 'That's a wonderful idea, Septic Barry.'

'I thought you'd like it. It's been a dream of mine for a few years now. And you know what? It's you who convinced me I had the potential to do it.'

She gave him a kiss. 'I'm so proud of you,' she said. 'Tell me all about it.'

He told her he would keep his usual business going while building up his new venture, *Contemporary Temporary Toilets*, on the side. 'It'll probably be quite slow to start with,' he said, 'but give me a few years and I'll be . . .' He scrunched his eyes shut, and gripped the bar top in concentration. 'I'll be . . . hold on . . . I'll be . . .' He took a deep breath, '. . . *one of the area's leading facilitators of short-term sewage solutions.*' He opened his eyes and relaxed his grip. 'That's going to be my slogan. I'm getting a truck with it written on the side.' Colour began to return to his fingernails. 'What do you think?'

'I like it,' she said. 'It makes you sound very modern. Very professional.'

'Actually, my dad thought it up. Anyway, I'm already booked for a silver anniversary in June, so I've got to pull my finger out.' He spoke about his plans to buy the equipment, and how that lunchtime he had met a man in a pub and agreed terms on a storage unit, and he recounted various details about the provision of temporary toilet facilities that Miyuki wished she could have gone through life without knowing.

She closed her book, nodded a goodnight and slipped out through the lounge bar door.

She had heard a lot of talk about the Gulf Stream, about how it keeps the area's weather mild. Sometimes it was possible to believe all that, but at this time of year, on a clear night after a clear day, it wasn't so easy. When she stepped outside it felt as if her eyeballs were turning to ice, and the breeze coming in from the harbour made her wonder how much colder it would have to get before the sea froze over. When she got in she put some kindling and a log on the embers, and it took a few minutes of rubbing her hands together for her fingers to come back to life.

She poured a glass of water and sat watching the fire. The log she had thrown into the stove was wedge-shaped, and looked like a big slice of chocolate cake. It seemed to be mocking her diet. Back at home Grindl made sure they ate steamed vegetables, fresh fruit, brown rice, yoghurt and high fibre cereals, but on these fortnights Miyuki allowed herself to be guided by cravings alone. If it hadn't been for Grindl, her regular weekly shop would have consisted of Pot Noodles, multi-packs of crisps, economy boxes of Jaffa Cakes, two-for-one deals on Pepsi Max, family-size tins of spaghetti hoops and crate after crate of the supermarket's own brand beer. Her liver would turn to sludge if Grindl ever left her, but she wouldn't give it a thought. She would let it turn to sludge.

Now that she had made it through an evening in the pub without being immobilised with a Taser gun and bundled into a police van, she thought again about Goldfinger. She was annoyed that she had been beaten to her idea. Painting the rock had seemed like such an original thing to do, and she was torn between wanting to kick whoever this person was, and wondering whether they were a kindred spirit. She wanted to know if Goldfinger was a run-of-the-mill vandal who just happened to have a bit of flair, or a pretentious artist making some kind of oblique statement about something of great importance, or maybe somebody like her who just wanted to turn a nice idea into something real, something that they and a bunch of strangers could enjoy. She didn't know what to feel, but whichever way she looked at it she knew that her rock-spraying days were over. At least she would be able to console herself with the knowledge that, Goldfinger or no Goldfinger, what she had done looked fantastic.

She could no longer think about the rock without thinking about tall Mr Hughes. She was finding it harder and harder to convince herself that he was just keeping a low profile. The time had come to find him. She decided she was going to track him down and listen to him for as long as he needed to talk. She would buy him a pint in The Anchor, and then everything would be fine.

She dipped her fingertips in the water, and plucked

out her contact lenses. They hissed and danced a samba, then opened and closed like a pair of transparent clams before lying still, alongside the remains of the others on the scorching metal.

She brushed her teeth, set her alarm and got into bed. She sniffed, sneezed her eighty-first unblessed sneeze, and picked up her photograph of Grindl. She gave it a kiss. She was missing her so badly she could feel it in her toenails and her kneecaps. She fell asleep to the sound of the wind as the apple sat on the worktop, forgotten in its paper bag.

SUNDAY

Gold glinted from the darkness as if nothing was amiss, but as the sun rose the light revealed patches of grey the size of potato prints, place-mats and dustbin lids. More than half the paint was gone. Piles of seaweed and pieces of driftwood lay inland from the rock, and Miyuki's cheeks burned in the cold air as she realised she knew even less about the sea than she had thought.

Kicking pebbles as she went, she walked to the other side of the cove and sat on the freezing stones, hugging her legs as she looked out to the horizon, down at her fingernails and over to the path, hoping somebody would come down and join her on the beach.

After an hour and a half she gave up.

She walked over to take one last look at the mess she had made, and saw that on the small rock beside the large one, apparently impervious to the assault of the waves, was the perfect golden print of the soles of a pair of size four Wellington boots. She stepped up to where she had stood before, half hoping that tall Mr Hughes would magically reappear, and half worrying that the police were going to be walking around the village looking at people's feet.

She stepped back down to beach level. It was time for her to go back to the cottage and start getting ready for church.

Miyuki always went to church on the Sunday in the middle of her stay. She told herself she was doing her bit for architectural conservation by keeping the numbers up so a nice old building wouldn't be closed down, demolished and replaced with holiday flats, but she had never been convinced by her own excuse. The truth was that she enjoyed herself. She liked listening to the hymns and the sermon, and soaking in an atmosphere that had hardly changed for centuries. She had grown up on streets teeming with churches and chapels but had rarely gone inside them, and even now she couldn't get a firm grasp on what distinguished one from the next, or why people had become so heated about the differences. On her first visit she had been drawn to the church because she liked the look of it, and was curious to see the inside. When she found she was able to sit quietly in the back pew she decided to make it a regular event.

Mr Puw arrived in a blazer and tie, his big black beard

violently combed into submission, and as he and his wife sat down a few rows in front of her, the organ started up.

She stood when she was supposed to stand, knelt when she was supposed to kneel, and sang along when she could. She recognised a couple of the hymns, but a lot of the time she hid her mouth behind the book and just listened to them. They sounded a bit disappointing this year. Without a rich baritone to bolster them, the voices seemed to get lost, drifting up to the high stone ceiling when usually they soared.

Halfway through the service she realised she wasn't doing what she normally did during the prayers. She wasn't peeping up to watch specks of dust in shafts of light, or trying to guess the average age of the congregation, or marvelling at the blackness of Mr Puw's hair. Instead her eyes were shut tight and she was thinking very hard about tall Mr Hughes, and hoping he was OK.

The collection tray came around and she gave three pounds, enough to buy Jesus a pint of beer and a packet of peanuts, but when the others went up for communion she stayed where she was. She didn't want to draw attention to herself by trying to copy them and getting it wrong. She wondered what would happen to her if she was to take communion without having the appropriate paperwork. She didn't want to spontaneously combust, or be eaten by locusts. She had a snack planned anyway, and didn't want to ruin her appetite by stuffing her face with wafers, so she was happy just to watch.

At the end of the service she filed out with everybody else. The late morning light shone through the bare branches of the trees and into her eyes, and as she passed the vicar she sneezed sneeze number eighty-four. She looked at him, and he looked at her, but he didn't say a word. He just smiled his vicar's smile.

She took her fried spaghetti sandwich to the bench in the back garden, and in between mouthfuls flicked pieces of bread on to the lawn for the birds. Before long high drama was being played out before her. For years she had watched these power struggles without really knowing who the participants were. It had been like coming late to a long-established soap opera, and one year, wanting to catch up, she had taken a spotter's guide with her. Now she was able to name every species that came to the garden, even the various small brown ones. She enjoyed the show until suddenly they all flew away, as if somebody had fired a gun. She looked up to see next door's cat on the garden wall.

'Furry bastard,' she mumbled, through a mouthful of pale food. Grindl would have been delighted to see it. She loved all cats, and Miyuki even had to put up with a framed photograph of one on the mantelpiece, an aloof black and white thing that belonged to her parents, and which would only ever raise itself from its torpor to go outside and commit casual genocide among the small creatures of the undergrowth.

With an empty lawn before her, she opened the Sunday edition of the *Star* that she had picked up on the way home from church, and found the photo story on the problem page. They were much as she had expected: a perky blonde and a sultry brunette, inexpensive models resplendent in lipstick and lingerie as they pulled poses for their exhausted storyline. The agony aunt went through the motions of offering her advice, urging the women to take care as they explored their feelings one step at a time. *Easier said than done*, thought Miyuki. She had never got the hang of treading carefully when it came to romance, and she had a feeling that if it hadn't been for Grindl she would still be out there, making the same mistakes over and over again. She could barely recall the names and the faces of the girls who had driven her out of her mind with love and sadness, but she could still feel the stab of their words as they told her they had been doing some thinking, and had come to a decision.

After scraping through her A-level retakes, Miyuki had got herself a place on a vaguely defined but broadly arts-based course at a former polytechnic. When she arrived she was delighted to find that the ultimate fashion accessory among a certain kind of student was a fluid approach to sexuality, and it wasn't long before she was pounced on by a girl who was keen to demonstrate to those around her that her sexuality was very fluid indeed. Night after night they lay entangled in the halls of residence as the girl delivered interminable monologues about how women understand each other's bodies so well, how life should be a journey of erotic discovery, and how her sexuality defied categorisation. 'I wouldn't say I was bisexual,' she droned, 'I'm just . . . *sexual*.' Miyuki found this a small price to pay, and as she half-listened to her lover's voice she luxuriated in the new feeling of somebody else's warm skin against her own. She hadn't wanted it to end, but it wasn't long before the girl decided her experiment was over, and allowed her fluid sexuality to guide her towards pleasures elsewhere.

Every term it was the same. She would be pounced on by a girl who went on to deliver interminable monologues about how women understand each other's bodies so well, how life should be a journey of erotic discovery, and how her sexuality defied categorisation. Miyuki desperately looked for the flashes of gold that told her she was in the right place, and whenever she caught a glimpse of one she clung to it like a prospector who's spent too long in the creek, frantically trying to convince herself that her life had been leading up to this, and that at last everything was going to go her way.

Miyuki's final year saw her romantic life back in its usual pattern. She was abandoned at the end of October by an Amazonian law student who for three weeks had pawed her in public places while looking around to make sure everybody saw, and the next term she found herself fleetingly coupled with an aspiring actress with corkscrew curls and enormous eyes.

By then she had learned to spot the warning signs, and to brace herself for the end, but the ability to anticipate the pain and humiliation didn't make it any less excruciating when it finally came. She would take the train back to her mother's, and wallow in a world of soup, Kleenex and words of comfort until she felt ready to take her bruises back out into the world.

When she got back after Easter she met a first year Biology student with a smile like the springtime, and fingers softer than she had ever thought possible, and things went the way they always did, right up until the moment she failed to vanish from her life. She was still there weeks later, and one Sunday afternoon when they were lying in bed, looking at each other with their feet touching, she said, 'I told my parents about you.'

'What did you do that for?' Miyuki snapped.

'Oh.' She had been hoping for a more or less opposite reaction to this. 'I just thought that we've been together for a while and, well, it's what you do, isn't it?' She was looking worried. 'Don't you like me? Is that it? I thought you liked me.'

'Yes, of course I like you.' She hadn't just been seeing the occasional flash of gold with this one, she had been seeing dazzling bursts of the stuff, a firework display. Right from the start she had begun bracing herself for the end, and she knew this fall would be the hardest yet. 'If you want to know the truth, I really *really* like you.' The moment she said this she wished she hadn't. The times she had declared her feelings were always coming back to haunt her. She was mortified to think that there were girls out there who knew, and who knew because she had told them, exactly how much they had meant to her.

'Then why can't I tell my parents about us?'

'Because you shouldn't be bothering your family with your *exploring-my-sexuality* phase. In a couple of weeks you'll be shagging a rugby player – you should tell them about him, not about me.' This was how it always ended, with the girl catapulting towards a straightforward brute, the affair with Miyuki becoming a footnote in her romantic history, affirmation that she was sexually adventurous, which was something she would want to know about herself before settling into a life absolutely lacking in sexual adventure.

'That's not true. I'm not going to be shagging a rugby player in a couple of weeks, I'm going to be shagging you.'

'Really?'

'Yes, really. So I'm afraid you're just going to have to get used to it.'

'But what about all that stuff you said about your fluid sexuality, and how you thought life should be a journey of erotic discovery?'

She turned red. 'Oh yes. God. Sorry about that.' She had delivered the same monologue as all the others. 'What a load of crap. Well, it's not *complete* crap – I do want to do a reasonable amount of erotic exploring, but only with you. And if you must know, my sexuality isn't as fluid as all that. In fact it's hardly fluid at all.'

'Are you sure about that?'

'I know it's hard to believe, but it's true, and that's partly why I told my parents. I had to tell them one day, and I thought it would be a good time, what with me having a steady girlfriend and all that.'

Miyuki was dazed by the thought that this girl saw her as a steady girlfriend, steady enough to tell her family. This conversation was a long way from the *big talk* she had been anticipating, and it was a while before she shuddered back to the moment and realised it was her turn to say something. 'So how did they take it?'

'Not too well.'

'Oh.'

'They've started reading the *Guardian* on Saturdays, so I thought they would be all *Guardian*y about it, and invite the neighbours round to celebrate with stuffed vine leaves and organic wine, but it turns out they only get it because the TV guide fits neatly on the side table. My mum burst into tears, and begged me to tell her where she had gone wrong, and my dad offered to send me to counselling. De-lesbianisation counselling, I suppose.'

'When do you start?'

'Very funny. Anyway, she kept saying I should get back with my ex-boyfriend. She always used to go on about how he was too old for me, and how I shouldn't trust a man with tattoos, but now she's completely changed her mind, and can't stop talking about how we should try and patch things up.'

'I think she's right, you know,' said Miyuki. 'You two should give it one last try.'

'Do you really think so? OK, I'll give him a call right now and see if he wants to take me to the Speedway for old time's sake.' She reached for her phone on the bedside table, and Miyuki wrestled it from her hand.

'Anyway,' said the girl, 'it all got out of hand after that. She was hysterical, and I ended up asking her why, if she was so keen for me not to go with girls, she gave me the name she did.'

'That's a fair question. What did she say to that?'

'She said, "Are you calling your great aunt Grindl a lesbian?" and I said, "Well, she never married, did she?" She said it was because she was happy on her own, and then she turned white. That was what the family had always said about great aunt Grindl, and nobody had ever questioned it, but it doesn't make sense, does it? Nobody's happier on their own. They might tell people they are, and they might even tell themselves they are, but they aren't. Not really.'

'So you're just like your great aunt Grindl.'

'I don't know about that. She died before I was born, so I never even met her. She might have been man mad for all I know. Anyway, I should have stopped there, but I didn't. I said that if I'd been called something normal, like Megan or Claire, I probably would have turned out all right. I told her that with a name like mine I'd never had a chance, just like great aunt Grindl had never had a chance.'

'You're a very bad daughter, you know.'

Grindl nodded. 'I knew I'd gone too far, and I kept telling her I was only trying to lighten the mood with a joke, but she took it to heart and now she really thinks it's all her fault. She said she would call me Megan or Claire from now on if that's what it would take to put things right. She even tried it once or twice. *Pass the pepper would you, Megan? So, Claire, have you met any nice handsome boys at college?*'

Miyuki could see there was a sadness to her smile, and she pulled her close.

'Anyway, I've told them now. And I had to tell them because I love you. There, I've said it. I've loved you since the moment I saw you. I wanted to slap that girl you were with – I knew you'd be happier with me, and I was right, wasn't I?'

Miyuki could only nod. It was a while before she could speak again, and when she did she said, 'I love you too.'

'Are you sure you mean that? A minute ago you said you just really, *really* liked me.'

'I was being coy. I'm really, *really* sure I mean it. I love you.'

'Thank God for that,' said Grindl. 'All this would have been a bit embarrassing for me otherwise, wouldn't it?'

Miyuki didn't know what to say. She knew she wasn't

going to be able to concentrate on her revision, and that she wouldn't get the marks she had been hoping for, but it didn't seem to matter. As she clung to the girl who loved her, it struck her that she would have to call home and tell her mother she would be visiting soon, and that there would be no need for her to stock up on soup and Kleenex because this time she would be taking somebody with her, somebody with a smile like the springtime, and fingers softer than she had ever thought possible.

She took her plate and newspaper indoors, and lay down for a five-minute rest. Three hours later she woke to find the daylight fading, and blinking her contact lenses back to life she put on her boots and coat and went down to the harbour.

She could see through their windows that The Boat and The Anchor were doing good business, but she resisted her body's many cries for beer and instead walked along the main street, looking in unlit windows at last year's menus, at oil and watercolour seascapes of variable quality, and at local pottery, love spoons and cuddly dragons, and reading the Blu-tacked signs that said things like CLOSED UNTIL MARCH, and STAFF WANTED END FEB – END OCT.

She tried to imagine what the village would be like in the summer, when these places were open and there were traffic jams and hordes of children everywhere, demanding ice cream and then smearing it all over their faces. She had never felt she was missing out by not being able to eat at the restaurants or browse in the souvenir shops, but she had thought about one day making a flying visit in the summer so she could take a boat trip out to look at the birds and the seals, and see the coastline from a whole new perspective. Maybe she would even take Grindl with her. Neither of them had ever seen a puffin, so that would be reason enough for them to make the trip, and she would enjoy showing Grindl around the place for the first time, and pointing out all the people she had told stories about. Grindl always asked after Septic Barry, and Miyuki had jokingly accused her of having a crush on him, an accusation that Grindl refused to confirm or deny. She pictured them having a pint together, then remembered that the way things were going she probably wouldn't be able to come back at all, with Grindl or without.

She stopped outside the shuttered booking office, and as she looked at the photographs on their notice board her heart thumped when she saw tall Mr Hughes looking back at her as he captained a small wooden boat full of smiling tourists in life jackets. For a moment it was almost as if she had found him, and didn't have to worry any more.

She looked at the photograph until a rumble from her belly sent her back towards the cottage.

She microwaved some tinned macaroni cheese and ate it on toast, with a bag of Frazzles as a side dish. This was one of her all-time favourite meals, but after another day with no tall Mr Hughes she hardly noticed she was eating it, and as she licked crumbs of processed maize from her fingers she felt a sharp throb behind her eyes.

She blew her nose, and saw that what came out was flecked with gold. She was in no doubt that this was an omen, but she couldn't tell if it was a good one or a bad one.

The afternoon drinkers had gone, and she arrived to find she was the only customer. She ordered a pint from Mr Edwards, and as she took it over to her usual place she glanced at the stuffed pike, trying to convince herself that it was giving her a look that suggested moral support. It wasn't long before short Mr Hughes came in, and a minute later Mr Puw. Mr Edwards served them their drinks before disappearing to his sanctuary behind the door marked PRIVATE. Standing at the bar, short Mr Hughes and Mr Puw exchanged some remarks about the weather and the quality of the beer, then lapsed into silence.

After they had been staring into space for a few minutes the refrigerator's thermostat clicked, and sent it rattling into life.

'That fridge needs looking at,' said short Mr Hughes.

Mr Puw nodded his agreement, and relit his pipe as the rumble filled the silence. It looked like being another night with no tall Mr Hughes, and Miyuki knew the time had come. She stood up, walked over to short Mr Hughes and Mr Puw, and took a deep breath.

Short Mr Hughes's mother had been a hot-blooded woman. If she ever caught her son misbehaving her face would turn scarlet, and with bulging eyes she would shout, 'I didn't bring you into this world so you could dip your dirty fingers into pots of freshly made plum jam,' or 'I didn't bring you into this world so you could throw pieces of gravel at a metal watering can while your father's upstairs trying to get some rest after a hard week's work,' or 'I didn't bring you into this world so you could climb over the wall and chase next door's goose round and round their yard with a long stick of rhubarb in your hand.' This had happened with such frequency that he had never thought about what the words *I didn't bring you into this world* actually meant. To him they were just the sounds that preceded a very serious telling-off.

Her confession over, Miyuki stood before him as his face shone scarlet, his eyes bulged and he bellowed, 'I didn't bring you into this world so you could get up early on Thursday morning and go down to the cove to spray gold paint all over a large rock.'

'Holy mackerel,' said Mr Edwards, as the noise brought him out from behind the door marked PRIVATE.

'Steady now, Alun,' said Mr Puw.

But short Mr Hughes wasn't finished. His eyes still bulged, and remaining scarlet he roared, 'And I didn't give you the greatest gift of all – *The Gift of Life* – so you could throw it back in my face by sitting there night after night not saying a word about how you'd spoken to tall Mr Hughes on the beach that day.'

Miyuki had no idea what to make of this. Her vision blurred, and she felt a loss of balance. As she steadied herself something terrible happened: a tear ran down her face, and then another, and before she could do anything to stop herself she was sobbing into her hands. She hadn't cried in public for a very long time, and her technique wasn't everything it could have been. She tried to stifle the sobs, but ended up emitting a high-pitched whine that seemed to come from somewhere deep inside, and which she wasn't able to control.

It was Mr Puw who took command of the situation. 'Come along now, Thunderthighs,' he said, gently. Taking her elbow, he guided her to a stool at the nearest table. He got her beer and put it in front of her. 'Come and join us, Alun,' he said.

More pink now than scarlet, short Mr Hughes knew he had gone too far, and he sat down somewhat sheepishly, wondering what his mother would have said if she had known he had made a girl cry. 'Sorry about that,' he said. 'It was dishonourable of me, raising my voice at you.'

'It's OK,' said Miyuki, her crying now diminished to intermittent shudders. 'It's my fault. I should have told you earlier.' Short Mr Hughes and Mr Puw looked away as she blew her nose. 'The thing is, I didn't know if I was supposed to be worried about him or not, and I didn't want to poke my nose in where it didn't belong.'

'I understand,' said Mr Puw, looking at short Mr Hughes, who nodded his agreement. They waited until she was ready to carry on.

'Everybody has bad days every once in a while,' said Mr Puw, when Miyuki had finished her summary of their meeting by the rock. 'I expect he's just gone to stay with a friend or a relative to cheer himself up.'

'Where do his family live?' she asked. 'Are they in the village?'

Short Mr Hughes and Mr Puw looked at each other. 'Up the coast somewhere, isn't it?' said short Mr Hughes.

'Either up or down, I think,' said Mr Puw. 'One or the other, anyway.'

'Does he ever mention anybody?' she asked, surprised that they seemed to know so little about someone with whom they spent so many hours of their lives. 'Was there ever a tall Mrs Hughes?'

Short Mr Hughes scratched his head, and said, 'I think I remember him saying there had been somebody a long time ago. A fiancée, I think it was. Something along those lines, anyway. That would have been before he came to live here, though. That's all I know. You don't ask about things like that, do you?'

'Don't you?' Miyuki really had no idea. There was a

silence, and she couldn't stop herself from asking something she had been wondering about for days. Whether she was supposed to ask such a question or not, she addressed them both. 'Do you like tall Mr Hughes?'

They looked at the table. 'We like him,' said short Mr Hughes. He looked at Mr Puw. 'Don't we?'

'We do.' Mr Puw nodded. 'We do.'

They sat in silence for a while, then short Mr Hughes spoke. 'Here's an idea,' he said. 'We could phone the oven chips people and ask if they can put us in touch with that dwarf. Maybe he'll know where he's got to.'

This was greeted with silence, and short Mr Hughes let the suggestion drop.

Mr Puw at last lifted his eyes from the table. He looked at Miyuki. 'Tell you what, Thunderthighs – drink up, and we'll all go for a walk.'

They took a narrow road up from the harbour to a part of the village that Miyuki had only visited once or twice. There wasn't much to see apart from a few streets of post-war houses and a phone box. They got to the end of a cul-de-sac, and Mr Puw rang the bell of a semi-detached house. In the moonlight she could see that tall Mr Hughes looked after his half of the pebbledashed building, but the other half seemed abandoned, with peeling paintwork and yellowing curtains, closed behind filthy windows. The front lawn was scraggy and knee-high, and a moss-covered FOR SALE sign leaned sideways from the hedge. Mr Puw rang again, and knocked, and there was still no response.

'He must be away somewhere,' said Mr Puw. 'Never mind. Since we're here, Thunderthighs, come and have a look at this.' They walked a few paces to the side of the house, and as short Mr Hughes lit the scene with his torch she saw that sitting on a trailer under a lean-to was a boat made from dark wood. The same boat, she supposed, as the one in the photo.

'He built her from scratch, you know. He's built plenty of others too in his time, but this one he made for himself. He looks after her, too. You wouldn't think she was getting on for twenty years old, would you?'

Even though Miyuki knew almost nothing about boats she knew a bit about woodwork, and she could tell it was a superb piece of craftsmanship.

Mr Puw patted the boat. 'He gets up to all sorts in this in the summer,' he said. 'He's always out there up to something. Trips round the islands, fishing, fixing up other people's boats. If you want a good job done around here you go straight to tall Mr Hughes. You don't go to the short one. Bloody hell, you'd sink.'

Short Mr Hughes smiled at this. 'It's true – you'd sink all right.'

Miyuki had seen Mr Puw out on his rounds often enough, but she had never thought to wonder whether either of the Mr Hugheses had a life beyond their evening drinks in The Anchor, their vegetable patches and their walks around the local footpaths. To her they had been just a pair of old men. Seeing a photograph of tall Mr Hughes taking tourists out in a boat had been a surprise, but finding out he was capable of building a boat from scratch, and that he was highly regarded in his trade, made her feel ashamed for having made up her mind about him so completely, and so lazily.

As short Mr Hughes shone his torch around she noticed a small window at the side of the house. It probably belonged to a downstairs toilet, and it was very slightly ajar.

The torchlight snapped off, and none of them said a word as they walked back to The Anchor.

She opened her book for the first time. It was the shortest one she had brought with her, but even so she wasn't sure she would be able to get through it that day. She was pleased to find she didn't like it very much, and didn't feel too bad about skimming over some of the descriptive passages. Towards the end of Chapter Seven a shadow fell on the page, and she looked up to see short Mr Hughes and Mr Puw standing over her with their coats and caps on.

'We'll be off now, Thunderthighs,' said Mr Puw. 'And don't you worry about tall Mr Hughes. He'll turn up.'

Miyuki nodded.

'And don't worry about that business with the paint either,' said short Mr Hughes. 'We won't go telling on you, as long as you promise not to do it again.'

'I promise,' she said, feeling more like a naughty child than a grown woman. 'I really wish I hadn't done it, you know.'

'We know that,' said Mr Puw. 'We all make mistakes, Thunderthighs. Even he's made one or two in his time, believe it or not.'

He pointed a thumb in the direction of short Mr Hughes, who nodded his agreement, and said, 'I went to see it this afternoon, as a matter of fact. It looks bloody horrible.'

'I know,' sighed Miyuki. 'I know it does.'

'It must be a punishment in itself for you, seeing it in that state,' he carried on. 'I don't think there's any need for us to call the police – you must have suffered enough out of sheer embarrassment.'

Miyuki nodded.

'Still, nice try, Thunderthighs,' said Mr Puw. 'See you tomorrow night.'

She nodded, and watched them go away.

A while later a man with a very blond beard came in through the PUBLIC BAR door. Septic Barry stood up to greet him, and took him to one side, away from the Children from Previous Relationships. As they stood at the far end of the bar they quietly engaged in what looked to Miyuki like a very serious conversation. So serious that she supposed it could only have been on the subject of human urine and faeces.

She went back to her book, and was thinking she might

even be able to finish it when Septic Barry left his drinking partner and came over to sit beside her. He told her all about his plans for Contemporary Temporary Toilets, and she listened for a second time to a series of unnecessarily graphic details about the ins and outs of short-term human waste management. When he was finished she wished him luck. They sat in silence until Septic Barry said, 'Seen him yet?'

'Not yet. Have you?'

He shook his head. 'He'll turn up.' With that he said goodnight, and went back to his caravan.

Miyuki finished her beer, took her glass back to the bar and followed him out into the cold.

She stoked the fire, poured a glass of water and took out her contact lenses. They did some quite impressive salsa dancing before giving up and lying still, looking like a pair of cellophane taco shells. She couldn't get the sorry state of the rock out of her mind. Short Mr Hughes had been right – she squirmed with embarrassment. She consoled herself with the thought that so much paint had been wiped off in such a short time that it wouldn't be long before it was all gone, as if she had never been near it with a spray can. She wished she had never done it, but even more than that she wished she hadn't gone back to look at it after the evening of its first day, and that she could have remembered it only as it had been when it had shone so beautifully. If she had never gone back it would have stayed perfect forever, and it really would have seemed as if she had given a gift to the world.

Watching the blurred flames through the stove's glass door, she couldn't stop herself from thinking about the news stories that made the papers every once in a while about people being found in front of flickering televisions after months, or even years, because nobody had wanted to poke their nose in where it didn't belong. She saw herself sitting in The Anchor in a year's time, with short Mr Hughes and Mr Puw standing at the bar, staring into space and saying, *He's probably visiting relatives, or something*, and *He'll be in tomorrow – he's not going to miss three-hundred-and-sixty-nine days in a row*. She pictured Septic Barry coming over to sit beside her, and asking if she's seen him around, and when she tells him she hasn't, saying, *He'll turn up*.

She wondered if the time had come for her to poke her nose in where it didn't belong.

She made one last attempt to get to the end of her book, but it hadn't got any better and after a few lines her eyelids drooped. Defeated for the first time in eight years, her head fell sideways and the book slipped from her lap on to the floor.

MONDAY

Miyuki had been right about the window. As she pulled herself up and shone the torch in she could see it belonged to the downstairs toilet. Her boots scraped against the pebbledash as she climbed through head first, trying hard not to put her knees through the glass. She steadied herself on the cistern as she eased her way down. It seemed to be even colder inside the house than outside, and her hands were shaking in their gloves as she untangled her limbs and turned herself the right way round.

On the walk over she had thought through exactly what she needed to do. Determined not to waste any time, or to give herself a chance to change her mind and climb straight back out of the window, she stepped into the hallway. The kitchen door was open, so she started there.

The torchlight revealed an old electric cooker, a fridge and a length of Formica worktop. Apart from a pile of jigsaw boxes on the table, everything in the room was functional. It took her just a few seconds to shine the light into the dark corners. She turned around and went back along the hallway.

No matter how softly she trod the sound of every footstep was almost deafening, and she wished she had worn trainers instead of boots. She pushed open the door to the living room, where an armchair sat in front of a television and three library books about alligators lay on the bare floorboards. There was nothing more to see, apart from a pile of logs and kindling beside a wood stove. She walked over, took off a glove and touched the metal. It was freezing cold.

Relieved to have got half the house out of the way already, she went to the foot of the stairs.

'Mr Hughes,' she called. There had been no response when she rang the bell, and she hadn't been expecting one this time. Remembering that the house next door was empty, she raised her voice. 'Tall Mr Hughes . . .' She wished she had planned what to say. 'It's Miyuki . . . I've come to see if you're all right.' She gripped the banister with her free hand as the words reverberated around the bare walls. 'I've been worried about you.' She waited for a moment, and made her way up.

Three closed doors led off the landing. She tried her hardest not to think about what might be waiting for her behind one of them, but it was no use. Terrible images that she had managed to suppress flashed across her mind, and as she banished each one it would be replaced by another just as awful, or even worse. As she shone the torch around she was acutely aware of the darkness behind her.

Not wanting to drag things out for a moment longer than she had to, she opened the door nearest to her and found herself in an avocado bathroom, with a greying net curtain at the window and embossed tiles on the walls. A toothbrush, a tube of toothpaste and a razor stood in a mug on the sink, and the air was thick with a miasma of cheap unscented soap. It was like a thousand bathrooms she had redecorated, and for a moment it felt strange that she wasn't there to tear down the tiles, rip out the suite, pull up the lino and throw a little piece of unwanted history into a skip.

She returned to the landing and tried the door at the back of the house. It opened into a small room with plain walls and bare boards. The only piece of furniture was a narrow bed that couldn't have been more than four feet long. She closed the door and turned around.

She could feel her composure draining through the soles of her boots as she walked towards the remaining door, and gently knocked. 'Mr Hughes,' she said. 'Tall Mr Hughes, it's Miyuki.' She tried not to think as she turned the handle and pushed it open. It was a large room that seemed even larger for its lack of contents. There was just a single bed, a chest of drawers and a big wardrobe. She tried her hardest to feel as though she had done what she came to do, but she couldn't take her eyes off the wardrobe.

She walked across the room. Holding her breath, she took the handle and opened it up.

Inside was a coat on a hanger, and a scuffed pair of work boots. Apart from that, it was empty. Breathing again, she pushed the door shut and looked around. The furniture was simple and solid, and she guessed tall Mr Hughes had made it all himself. Compared to the stark minimalism of the rest of the house, this room was almost opulent. There was a rug and a bedside lamp, and on top of the chest of drawers stood a framed black and white photograph, the only real adornment in the house. She couldn't resist picking it up for a close look. A young woman smiled blankly as she stared at the wall. Whichever way Miyuki held the frame the woman seemed to be avoiding eye contact. She was wondering whether she had found somebody with whom she could share at least some of the blame, when the doorbell rang.

She almost crushed the frame in her hand as her heart tried to fight its way up her windpipe. *Oh Jesus*, she thought. A moment later there was another ring, and a knock, and before anybody would have had a chance to answer, a key rattled and clicked in the lock. She looked around for an escape route, but there wasn't one. There was only one place she could go, and still holding the photograph, she opened the wardrobe and got inside as quickly and as quietly as she could. She switched off her torch and pulled the door shut as she folded herself into a foetal position.

She could hear footsteps on the floorboards downstairs. A man cleared his throat, then slowly made his way upstairs and on to the landing. The bedroom door was still open, and he stepped inside. He switched on the light, and as it shone through cracks in the wood she tried her hardest not to move a muscle. She could see her hands shaking, and worried that if she trembled any more the wardrobe would start shuffling across the floor like a clockwork toy. She tried to empty her mind, but couldn't.

Tall Mr Hughes wouldn't have rung his own doorbell, so she knew it couldn't be him. As her thoughts fell into place she realised that the house next door couldn't have been empty after all. Whoever lived there must have heard her scraping her way up the pebbledash and calling up the stairs, and come to investigate. She and Grindl had exchanged spare keys with their next-door neighbours, so there was no reason why tall Mr Hughes wouldn't have done the same with his. The difference was that their neighbours were a friendly old couple, with house plants on the window sill and a shiny front door. She wasn't impatient to be found hiding in a cupboard by the kind of person who lived in a house like the one on the other side of the bedroom wall, the kind of house that could only be lived in by somebody who spent his days in self-imposed darkness doing little besides eating live woodlice, painting his toenails, drinking his own urine from a chipped gravy boat, masturbating into his dead sister's wedding dress and sharpening his enormous collection of knives in readiness for a moment just like this.

He stood there for a while longer, his shoes creaking as he rocked backwards and forwards. Miyuki could just make out the photograph. The woman was looking straight at her, and Miyuki was sure her eyes had narrowed and a sneer had crept into her smile. She turned the frame over. The creaking carried on, and she awaited the inevitable. She wanted to cry out for her mother, and for Grindl.

The inevitable didn't happen. He switched off the light and left the room, closing the door after him. Back in pitch darkness she allowed herself to exhale at this anti-climax, then listened as he opened and closed the other upstairs doors.

The footsteps came back along the landing, and stopped outside the bedroom. She willed him to go downstairs, but it didn't work. Instead he came back into the room, switched on the light, walked straight over to the wardrobe and opened the door.

When she was growing up, Miyuki had gone through phases of wondering what her name meant. One morning, over Rice Krispies and out of the blue, her mother told her she had been named after somebody who had shown her great kindness in Osaka. She hadn't volunteered the nature of this kindness, and Miyuki hadn't asked her to explain. They finished their breakfast to the sound of the cereal. Another time, again at the breakfast table, her mother had said, 'If you ever want to know about your dad just ask, and I'll tell you everything I can. But I'm afraid it won't be very much.'

Miyuki had always found a reason not to ask. Her mother hardly ever spoke about her time in Japan, and whenever she did Miyuki saw a sadness in her eyes, as if she had left something precious behind, something she knew she could never get back. This would have been reason enough for her to let the subject of her name drop, but what really kept her from asking for more details was the nagging worry that it might mean something embarrassing, and that it would be best if she never found out.

One afternoon, though, she had found herself in the college library after a couple of drinks, and was gripped by a sudden and fearless curiosity. She hurled herself into some furious cross-referencing until she found the book she needed. When she reached her name, she read the meaning over and over again:

The silence of deep snow.

At first she had no idea what to think, but after some deliberation she concluded that it could have been a lot worse, and went back to the pub to celebrate.

Since then she had more or less been at peace with her name, and it only ever grated when she found herself living up to it. Sometimes, in situations where she knew she ought to be saying something, her mind would turn brilliant white. It wasn't so much that she couldn't find the right words to say, the problem was that she couldn't find any words at all, as if they were hiding from her. Sometimes this lasted just a few seconds, sometimes a minute or two, and sometimes even longer, and as soon as she came to she would curse the woman from Osaka who had been so kind to her mother. If her name had meant something more practical, like *Hail on a tin roof*, she wouldn't have found herself in these situations. She could have been more like Grindl, who was never stuck for something to say. Her problem had crept up on her in class, and in job interviews, and when she had found herself alone with Grindl's parents, and on the beach with tall Mr Hughes.

It was happening again as she looked up in silence from the bottom of the wardrobe.

The man's face turned from white to scarlet by degrees, his eyes bulged, and he bellowed, 'I didn't bring you into this world . . .' before spluttering to a halt. He had another try, and this time his voice was higher. 'I didn't give you the greatest gift of all . . .' Again he stalled, and Miyuki heard somebody else come bounding up the stairs.

'Morning, Thunderthighs,' said Mr Puw, as he appeared in the doorway. 'You're up early.'

Her mind remained brilliant white as a tear ran down

her cheek, and then another, and her face crumpled as she sobbed into her hands.

'Here we go again,' said Mr Puw. He left the room and came back with a length of toilet paper.

She blew her nose and dried her eyes, and grasped through the blankness for some words that would explain what she was doing at the bottom of tall Mr Hughes's wardrobe. All she could manage was, 'I just wanted to know if he was here.'

'Great minds think alike, Thunderthighs. Great minds think alike.'

'So you *are* worried about him?' she asked.

'Oh no, we're not worried about old Hughesy. He'll be fine.'

'Then why . . .' Miyuki looked baffled.

'Well, maybe we're just a little bit worried, if you absolutely must know. We'd all swapped spare keys a few years back, so we thought we'd have a quick check before breakfast. Just in case. He's got a workshop down by the harbour and we had a look in there last night too. But don't you worry about him, Thunderthighs. He's probably visiting relatives or something.'

Miyuki nodded. 'Probably.'

'So,' said short Mr Hughes, his face less scarlet than it had been, but still quite red. 'Vandalism, housebreaking . . . any more strings to your bow we ought to know about?'

'Er, that's about it when it comes to crime,' she said. 'But I'm getting better and better at cooking crumble.'

'Let's be on our way,' said Mr Puw. He offered her his hand, helped her out of the wardrobe and led the way out of the room. On the way past the chest of drawers Miyuki put the photograph back in its place, and took one last look at the woman's face. There was no sneer, and her eyes weren't even slightly narrowed. In fact she looked quite pleasant.

They left through the front door. Short Mr Hughes went past the boat and had a quick look around the back garden. He came back shaking his head, and they walked away. They didn't see anybody else until they got to the harbour, where a figure was slowly writhing on the quay, in the dull yellow light of the single street lamp.

'Look at that bloody idiot,' said Mr Puw, affectionately.

They stood in a row and watched him for a while. When they had seen enough they went their separate ways.

When she woke up it was almost noon. After four slices of toast and margarine and a cup of tea, she gathered her things and went out into another bright day. A few miles along the path she found a comfortable rock at the end of a promontory.

A boat was heading her way, going in and out of view as it hugged the shoreline. As it came close she could see it was a lifeboat, and when it sailed past she recognised one of the crew as the blond man Septic Barry had been talking to in The Anchor. Her eyeballs throbbed. She had blandly assumed that their conversation had been about sucking shit, piss and paper up a pipe. She felt like an idiot, and wondered if she was ever going to stop jumping to conclusions.

She relived yet again the last time she had seen tall Mr Hughes. It was the only real conversation they had ever had, and she had made a mess of it. Banging a clenched fist against her forehead, she wondered whether she should apply to join The Samaritans. She could get angry with her callers, direct them to some handy suicide spots, and at the last minute feel guilty and offer to buy them a beer and advise them to do something fun. That would help them out.

As her bones cried out for Grindl she was hit by the realisation that their separation was a stupid idea. The times they spent apart as they went about their everyday lives were more than enough to get the result she was after. Whenever Grindl was spending a few hours in the spare room on the computer, or visiting friends for the weekend, or busy with something to do with the youth club, Miyuki missed her. When she held her as she slept, and heard her rumbling breath and felt the blood as it pulsed around her body she could sense time slipping away from them, and she knew without having to think about it that she would never take her for granted. She was like a self-made millionaire taking a journey by bus to remind herself of the way things used to be, but no matter how hard she tried it would never be like the old days, not really. If she kept chasing this synthesised approximation of melancholy it wouldn't be long before they had spent a whole year apart, and for no good reason. She felt winded, and wanted to run to the nearest phone box and call home, and tell Grindl she didn't want to do it any more, that she would be back in the morning, but she couldn't. Not until there was some news about tall Mr Hughes.

She got up and headed back. As the village came into sight she sneezed her ninety-fourth consecutive unblessed sneeze. She remembered she had the apple with her, and took it from her pocket, eating it as she went along. As she dropped the core into some long grass, she sneezed again.

It hadn't worked.

When she woke up for the third time that day she showered, ate a bowl of tinned custard and a flapjack, and rummaged under the sink for a black bin-liner. She took it through to the bedroom, where she filled it with empty spray cans and tied it closed before taking it outside, dropping it into the dustbin and putting a bag of kitchen waste on top. She told herself she would plant a tree in the back yard when she got home, to make up for the pollution she had caused. She thought of all the trouble she had caused on this trip, and wondered whether she should plant an entire forest to make up for it.

She went back in and got ready for the pub, picking up the boring book from the day before and putting it in her coat pocket. She doubted she would even open it that day, let alone finish it. She put a log in the stove, and went down to The Anchor.

To the dismay of the regular contestants, four walkers from Caerphilly who were staying in the rooms upstairs had decided to join the quiz. They called their team The Four Walkers from Caerphilly, and seemed unnervingly keen as they huddled around the table in between the dartboard and the cigarette machine.

Septic Barry's bass player was handling their paperwork. When quiz night first started they had changed their team's name every week. To begin with these had been fairly innocuous, The Last Living Pandas one week, Shave the Children the next, and Can You Hear My Hands? the week after that, but it didn't take long for Septic Barry to realise there was a seam of mischief waiting to be tapped. One summer evening they called their team Look Out! Wasp!, and were delighted to see that this caused mild alarm and darting eyeballs whenever the scores were read out at the end of a round. The week after that they named themselves Ladies and Gentlemen, I Regret to Announce the Passing of Prince Charles, which resulted in bowed heads and a respectful hush. This went on until the evening they called their team There Has Been a Bomb Scare – Please Move Slowly and Quietly Outside. This Is Not a Drill. Leave Your Personal Possessions and Vacate the Building Immediately. On hearing this, a pair of nervous young birdwatchers leapt to their feet and charged outside to safety, refusing to re-enter the pub until there had been an official all-clear. Mr Edwards took Septic Barry aside, and his *Holy mackerel* told him that enough was enough. Septic Barry saw his point, and from that night on their team went by the name God had given them: Septic Barry and the Children from Previous Relationships. His bass player wrote this at the top of the sheet and listed their names, their instrument in brackets after each one.

Things weren't so straightforward in the lounge bar. Short Mr Hughes and Mr Puw's team had always been called Hughes Puw Hughes, but without tall Mr Hughes it didn't seem right. After a brief conference they decided to show there were no hard feelings about earlier by calling themselves Hughes Puw Japanese Girl. Short Mr Hughes had just finished writing this at the top of their sheet when he looked up to see the lounge bar's door open, and tacked a second Hughes on the end.

'Just in time,' said Mr Puw.

'Got your pound?' asked short Mr Hughes.

Tall Mr Hughes reached into his pocket, pulled out a coin and added it to the pile on the bar. 'Must be my round by now,' he said. He nodded to Mr Edwards, who nodded back and pulled three pints of Brains without having to be asked.

A horrible screech of feedback told them that the quiz was about to start, and they took their usual seats at the table by the fire. After another brief conference, Mr Puw turned to Miyuki.

'Thunderthighs,' he called. 'Why don't you join us?'

She had been hoping he would say that. She gathered up her book, coat and beer, and slid along the bench.

The second question in the third round was, *What is the name of the largest of the Japanese islands-ah?* This elicited a loud groan from the public bar, and of course there was no surprise in the lounge bar that their team member knew the answer. What did surprise them, though, was that she knew the answers to quite a few of the other questions too, questions that had little or nothing to do with east Asia. She knew the names of Henry VIII's wives, in order, the boiling point of alcohol, the exact height of Pen y Fan, and the currencies of countries that couldn't have been further from Japan. Round by round their combined knowledge gave them an increasingly comfortable lead over The Four Walkers from Caerphilly.

It wasn't until Hughes Puw Japanese Girl Hughes were drinking their winners' pints that Mr Puw turned to tall Mr Hughes and asked, 'Been away?'

'I have been precisely that.'

'Where to?'

Tall Mr Hughes took a deep breath, paused for effect, and said, 'I have been to . . . the Wiltshire town of Trowbridge.'

Short Mr Hughes and Mr Puw almost choked on their

Brains. 'Trowbridge?' spluttered Mr Puw. 'What were you doing in bloody Trowbridge?'

'Well, I got talking to this chap in Milford Haven.'

'Milford bloody Haven?' squawked short Mr Hughes. 'At this time of year?' Tall Mr Hughes had long expounded the belief that although the county's other towns were acceptable destinations all year round, it was pointless to go to Milford Haven before Pancake Day. Nobody could quite follow his logic, but he seemed to have a good reason for this, and it was a subject he would return to with some frequency.

'Just taking a break from routine,' he said. 'Anyway, this chap stopped me in the street and asked if I knew the way to the railway station, and since I had nothing else to do I walked with him. I helped him on with his bags, and I was only halfway through saying my good-byes when the train pulled away. Then the ticket inspector came along and I didn't know what to do, so I had a think about it and ended up deciding to have an adventure. I showed her all my money and asked how far it would take me. And you know what she said?'

Short Mr Hughes and Mr Puw were apparently speechless, and Miyuki supposed one of them ought to say something. '"Trowbridge?"' she ventured.

'Precisely. So that's where I went. One change at Newport, then across the border and onwards to Trowbridge. Total journey time: four hours and thirty-nine minutes.'

'And what did you do when you got there?' asked Mr Puw.

'First of all, since I was penniless, I withdrew some money with my cash card.'

'Cash card? Since when have you had a bloody cash card?'

'Since 1987.'

Short Mr Hughes and Mr Puw exchanged incredulous glances.

'First time I've ever used one, mind.'

This seemed to restore some kind of balance to the world, and short Mr Hughes ran his fingers over his moustache while Mr Puw relit his pipe.

'So I found a room in a B.&B., and for the next few days I followed the town's many trails, looked around the garden centre, went to the pub, paid my respects at the war memorial . . . Things like that, really.'

Neither Mr Puw nor short Mr Hughes felt the need to ask any more questions. They knew tall Mr Hughes would be talking about his trip for days to come, maybe even weeks. It would be *Trowbridge* this, and *Trowbridge* that, and although they would never have admitted it they were just slightly looking forward to hearing a bit more about what he had been up to.

'Nice place, Trowbridge,' he said.

Seeing their glasses were nearly empty, Miyuki went over to the bar to order a round. She had been beaten to it by the four walkers from Caerphilly, and as she waited her turn she heard tall Mr Hughes start up again, his voice lowered as if in confidence, but still loud enough to be heard across the quiet bar.

'And do you know who I have to thank for my trip?' he asked. Short Mr Hughes and Mr Puw shook their heads. 'Our little Japanese friend over there. Miyuki, her name is.'

She pretended she hadn't heard, and carried on looking away as Septic Barry's girlfriend pulled the walkers' pints.

'It's true,' he carried on, apparently still under the impression that he was whispering. 'I met her on the beach on Thursday morning. She was covering a rock with gold paint, by the light of the welkin.'

Short Mr Hughes and Mr Puw waited for him to carry on.

'I suppose you're wondering what *welkin* means, aren't you?'

Neither of them was wondering any such thing, and they stared at their glasses as its uninvited definition washed over them.

'It looked quite nice as a matter of fact. Not entirely to my taste, but a good job nonetheless. Anyway, I'm afraid to say I was somewhat despondent that day, and do you know what she said to me? She said, "Tall Mr Hughes, you have to follow your dreams. The world is your oyster, you should go out and look over the horizon and see what lies beyond. Who knows what wonders you will find?" All that sort of thing. It really struck a chord, and by the middle of the afternoon there I was – halfway round Trowbridge's historical Town Trail.'

Miyuki felt herself filling up with pride. She decided not to remind tall Mr Hughes of her exact words – all that mattered was that she hadn't done such a bad job of cheering him up after all. Pretending she hadn't heard a thing, she ferried the drinks over to her team mates, one at a time.

'I was just telling this lot,' said tall Mr Hughes, as she sat down, 'that you really lifted my spirits that morning. I expect you noticed I was somewhat downcast at the time. It's not like me to get like that, but there it is. And talking to you snapped me right out of it.'

They all looked at her, and she couldn't think what to say. 'Well,' she started, '. . . everybody has bad days every once in a while.' They carried on looking, clearly waiting for her to carry on. 'And besides . . .' Before her mind had a chance to go blank she clutched at the first phrase that came to mind. '. . . the past is another country.'

Mr Puw chuckled, and shook his head. 'There you go again, Thunderthighs,' he said. 'Dispensing your pearls of Oriental wisdom.'

'You're right, you know,' said short Mr Hughes, looking in the direction of the pewter mugs hanging behind the bar. 'The past *is* another country . . . and tomorrow is another day.'

They all nodded their agreement. 'On that note,' said

Mr Puw, 'may I take this opportunity to propose a toast?' He raised his glass. 'To . . . what did you say your name was?'

'Miyuki.'

'Japanese, is it?'

She nodded.

'I thought as much. Anyway, here's to . . . what was it again, Thunderthighs?'

Nobody noticed it happening, but at the time when he would normally be clearing his throat as a gentle hint for people to start making their way home, Mr Edwards drew the curtains and put another log on the fire. He raised a hand in goodbye as the quizmaster staggered into the night, nodded to the four walkers from Caerphilly as they headed up to bed, and bolted the front door.

It was going to be a long night, and everybody had already drunk more than usual for a Monday.

'Married yet, Thunderthighs?' Mr Puw asked Miyuki.

'No,' she said. 'Not yet.'

'Oh well. Don't give up hope – I've seen uglier girls than you find a husband.'

Tall Mr Hughes and short Mr Hughes nodded their agreement. Miyuki knew that Mr Puw's catastrophically chosen words were meant with kindness, and she said, 'That's reassuring to know.'

'How old are you, anyway?' he asked.

'Bloody hell, Bryn,' said short Mr Hughes. 'You can't

ask a lady her age. You'll be asking her how much she weighs next.'

'Oh, I'm sorry, Thunderthighs. It's the Brains talking. I forgot my manners there. Well interjected, short Mr Hughes. Let's pretend I never asked.'

'No, I don't mind telling,' she said. 'I'm thirty-one. Thirty-two soon.'

Mr Puw looked up at the ceiling, and smiled. 'My daughter would have been about your age by now. You'd have liked her, Thunderthighs.'

She didn't know what to say, so she just nodded. They sat in silence until it was broken by tall Mr Hughes.

'Just out of interest,' he said, turning to Miyuki, 'how much *do* you weigh?'

Short Mr Hughes and Mr Puw shook their heads, but Miyuki smiled and told him. At that moment she would have answered anything, as long as it was tall Mr Hughes who was doing the asking.

As her team mates' conversation drifted towards the ins and outs of their vegetable patches, Miyuki excused herself and went over to sit with Septic Barry and the Children from Previous Relationships. Septic Barry's girlfriend was with them, drinking a glass of white wine and smoking a cigarette, and the Children from Previous Relationships seemed to be in a state of shock. 'Tell her then,' said Septic Barry's girlfriend.

'No, you tell her.' Septic Barry could see from the

Children from Previous Relationships' faces that a band meeting was in order. His girlfriend got the hint, and invited Miyuki to sit with her at another table.

Miyuki found out that Septic Barry's girlfriend had a name. She was called Siân, and Siân excitedly told her what they had just told the band – that she and Septic Barry were going to move in together.

'But you already live together, don't you?'

'I suppose we do, but I mean live together properly. In a house, like.'

'A house? Are you sure about that?'

'I know it'll be a change, but he'll get used to it. Me and my ex are selling our flat in Penarth, and my half of the money's coming through any minute, and with Septic Barry's savings on top of that we can afford to get a place round here. You know, it's amazing how much you can save if you never buy anything except baked beans, beer and guitar strings. We had a look at the place next to tall Mr Hughes's this morning, and had an offer accepted this afternoon. It needs a lick of paint, mind, but it'll be nice when it's done up, and it's got a spare room so we'll be ready when the time comes.'

Miyuki looked over at Septic Barry, who was wearing a happily dazed expression as he chaired the band meeting. Although he and Siân had only spoken of it euphemistically, he knew perfectly well what *when the time comes* meant. He had never thought that such a time would ever come, but as he pictured a miniature version of himself, or of Siân, following him around in a matching promotional sweatshirt in a very small size, it felt like something he was ready for, something he wanted to do.

Siân told Miyuki she was going for an interview to do her old job again, as a dental nurse at a nearby practice, and if she got that they wouldn't have to worry about paying the bills. 'Trouble is,' she smiled, 'I can't help thinking he's more excited about living next door to Hughesy than he is about living with me. He's been worried sick the last few days. I remember him telling me a while back that if he'd never had a father, tall Mr Hughes would have been the father he'd never had.'

Miyuki didn't know what to say. She looked over to the stuffed pike on the other side of the pub, and was sure it was looking back at her with an expression that said, *No, me neither.* Even after thirty-one years of hanging on the wall it hadn't come close to understanding the currents and the undercurrents that rippled around the room.

Over at Septic Barry and the Children from Previous Relationships' table, Septic Barry said, 'The band is finished . . .' Then, without warning, his eyes glazed over and his jaw went slack as he was floored by the realisation that even though he had done it countless times, he had never had sex in a building. To him sex smelled of grass, of the night air, and of sausages frying on Calor gas stoves. In the summer it was something done in virtual silence so nobody could hear through the canvas, and when he moved into his caravan for the winter it was always done huddled under a duvet to keep out the cold. Soon, though, he would be living in a house with brick walls and central heating, and sex would smell of a brand new carpet, and of freshly laundered sheets. His jaw slackened further still as he pictured himself and Siân living in their new home and *not* having sex. There they were, sitting on a sofa, talking and making each other laugh, and sometimes, just sometimes, spending whole evenings in because they didn't really feel like going to the pub that night.

The Children from Previous Relationships were pale with fear. Septic Barry snapped back to life, and finished his sentence. '. . . unless we take things up a gear. We can't go on this way. Tell you what . . .' He stood up and tapped his glass, and the pub fell quiet. 'I'm very sorry to interrupt,' he said, 'but I have a brief announcement to make. The Children from Previous Relationships and I will be performing a live concert over there, by the dartboard, on . . .' He looked over at Siân, then down at the floor. 'On my fiancée's . . .' He looked back at her and mouthed, 'Is that OK?'

She nodded.

'Anyway, what I'm saying is, we'll be playing by the dartboard on Siân's birthday, on the fifteenth of November.'

The pub remained silent as everybody absorbed the landmarks in this announcement. It was Mr Edwards who spoke first.

'Holy mackerel,' he said.

'You can say that again,' said Mr Puw. 'We'd all better get some ear plugs, then.'

A grinning Septic Barry sat back down.

'I just want to check,' Siân whispered to Miyuki, as the rumble of conversation returned to the room. 'Did I just get engaged?'

'I think you did. Congratulations.' She turned away, and ran a finger across the bottom of her left eye. She hoped nobody had noticed. She didn't want them to remember her as a cry-baby.

The Children from Previous Relationships became paler still at the realisation that they were going to have to learn some songs, and rehearse them, and play them in front of people with only ten months to prepare. This wasn't what they were used to at all.

'Well?' asked Septic Barry. 'Are you on board?'

There was a long pause, then the keyboard player spoke. 'I'm not sure,' he said. 'The trouble is, I don't think I'll be able to plug my keyboard in. There isn't a socket by the dartboard, see.'

'I've got an extension cable,' said Septic Barry.

The keyboard player thought for a while. 'OK then,' he sighed. 'I suppose I'll be up for it, as long as everybody else is.'

'So are you?' asked Septic Barry, turning to his drummer.

'Well,' he said, 'if the alternative is no more band then I don't suppose we've got much choice, have we?'

'That's the spirit. And how about you?' he asked his bass player.

'Do we *really* have to?'

'Yes, we do.'

'I might need an extension cable too, though. For my bass amp.'

'I think we'll be able to track one down between now and November.'

'Oh.' He sighed. 'OK then.'

'So that's decided,' said Septic Barry. 'We're going to rock this place like it's never been rocked before.'

The Children from Previous Relationships stared at their pints, and Septic Barry slipped back into a trance as he thought about how he would no longer be slinking around the campsite, sidling up to girls and leaning in to them, hoping they would lean back in to him. Never again would he suck a stranger's earlobes by the light of the moon. He looked over at his brand new fiancée, who was deep in conversation with the Japanese girl, and realised he wouldn't miss it at all. Not one bit.

'So when do you think you'll be getting married?' Miyuki asked Siân.

'I'll need to have a serious talk with Septic Barry about that. There's something very important we'll have to get out of the way before the big day.'

'What's that?'

'There'll be a deal. He's been on at me to give these things up,' she held up her cigarette. 'It's about time anyway, so I don't mind doing that. And he doesn't know it yet, but in return he'll be getting himself a haircut. If he thinks I'm walking up the aisle while he's got that thing on his head he's got another thing coming.'

Miyuki looked over at him, and could see Siân's point. She smiled, and said, 'All good things must come to an end.'

They were all still there at half past one, and were starting to flag when tall Mr Hughes spoke up, his rich baritone filling the room. 'There's this fellow in Trowbridge,' he said. The pub went quiet, and they all listened. 'Name of George.' There was a long pause, and everybody waited for what was coming next. 'Nice chap, he is.' Again they waited, wondering where this story was heading. It wasn't until tall Mr Hughes's eyes glazed over and he reached for his beer that they realised this was the extent of the anecdote. He had met a man in Trowbridge, the man's name was George, and this George character was nice. That was all.

Nobody said a word, but one by one they decided it was the best story they had ever heard. The pub came to life again, and Mr Edwards pulled nine pints of Brains and poured a glass of white wine, and refused to take a penny.

TUESDAY

At half past two in the afternoon, a pale Miyuki Woodward could be seen standing at the bus stop by the café, yawning into the back of her hand as she tried to make sense of the timetable. She had slept with her contact lenses in, and the whites of her eyes were pink as she squinted at the grid of numbers from behind a pair of National Health glasses. The next one was due any minute. There was no time to go to the phone box, so she rummaged in her big bag for her mobile. She hadn't looked at it since leaving home, and wondered whether the battery would be flat, but it started up and she put it in her coat pocket. It was another bright day, and as the sun moved out from behind a chimney she sneezed her ninety-eighth sneeze, blew her nose, and looked up to see the bus arriving.

She struggled on with her bags, and sat near the back.

The half-dozen other passengers seemed to be lost in their thoughts as they pulled away and rounded the bend into the harbour. A refrigerator repair van was parked on the yellow line outside The Anchor, and as the bus driver negotiated his way around it, Miyuki looked at the pub and thought about the night before. Then something happened.

She knew it was just a coincidence of light, and angles, and other things she didn't really understand, but it looked as if the whole building had turned to gold.

They got around the obstruction and carried slowly on, past the garage, where Septic Barry's van was parked outside as it waited for its service. His golden van. Everywhere she looked, something had turned to gold. An out-of-date menu in a restaurant window, the notice board outside the boat trip booking office, the pillar box, the awning over the door of the café, even people on the street, other passengers on the bus, and rooftops, trees and clouds. There was gold everywhere.

As they passed the last of the houses the bus picked up speed, and the flashes of gold became fewer and fewer until there were none. She bit off a chunk of fingernail, and surrendered to the realisation that she was leaving the village for the last time. As long as she never looked back, everything would stay just the way she had left it, bathed in a golden light.

She kept her eyes fixed on the grey road ahead as the village shrank behind her.

* * *

She took her phone from her pocket, and dialled. After a few rings the answerphone clicked on, and she cringed at the sound of her own voice as it delivered the GM Interiors spiel. Embarrassed to be breaking her rule about not making phone calls on public transport, she made a point of speaking quietly. After the tone she said, 'Grindl, it's me. I'm coming home, because . . . because I miss you, and I just want to come home.' There it was, the admission of defeat, and it hadn't hurt at all. 'Anyway, I've just got on the bus. I'm not sure about train times, so I don't know exactly when I'll be in. Some time this evening, anyway. I'll give you a call . . . Hold on . . .' She raised her hand to her face, and waited for number ninety-nine.

From her seat at the kitchen table, Grindl looked across the room to the answerphone. There was a silence from the tinny speaker, then a sneeze. She didn't say a word, and made no move to pick up the phone.

Sorry about that, the voice carried on. *Must be something wrong with the inside of my nose. I'm almost out of battery, but I'll give you a call from a box when I get a chance. I'll see you later, whatever happens. There's so much to tell you. I . . .*

The line went dead.

Grindl chewed the end of her pen for a while, then pressed her hand to her forehead, sighed, and carried on writing.

A cup of tea appeared on the table as she signed off and stuffed fourteen pages of neat, tiny writing into an envelope. She sealed it, and on the front she wrote, as though it were a greetings card, the single word, *Miyuki.*

Miyuki felt it first inside her left nostril, then it spread up to her eyes. She tried to stifle it, but it was no use. It was going to be a big one. Unstoppable. Not wanting her hundredth sneeze to be a muted snort, she let go of her nose, and it shook her bones as it blasted out.

One by one she looked at the other people on the bus, and none of them turned around. It was as if they hadn't heard a thing.

And then it was too late.

Slumping back in her seat, she looked through dusty glass at fields and houses, and caught glimpses of deep blue through gaps in the trees. She took off her glasses and put them in her pocket, and as the world slipped out of focus, a feeling of absolute tranquillity washed over her.

Resting her head against the window she closed her eyes, and waited for her heart to stop beating.